Advances in Intelligent Systems and Computing

Volume 234

Series Editor

J. Kacprzyk, Warsaw, Poland

For further volumes:
http://www.springer.com/series/11156

Advances in Intelligent Systems and Computing

Volume 234

Series Editor

J. Kacprzyk, Warsaw, Poland

For further volumes:
http://www.springer.com/series/11156

Jeffrey W. Tweedale · Lakhmi C. Jain
Editors

Recent Advances in Knowledge-based Paradigms and Applications

Enhanced Applications Using Hybrid Artificial Intelligence Techniques

Editors
Jeffrey W. Tweedale
Aerospace Division
Defence Science and Technology Organisation
Edinburgh
Australia

Lakhmi C. Jain
Faculty of Education, Science, Technology and Mathematics
University of Canberra
Canberra
Australia

ISSN 2194-5357 ISSN 2194-5365 (electronic)
ISBN 978-3-319-01648-1 ISBN 978-3-319-01649-8 (eBook)
DOI 10.1007/978-3-319-01649-8
Springer Cham Heidelberg New York Dordrecht London

Library of Congress Control Number: 2013948740

Printed on acid-free paper

Springer is part of Springer Science+Business Media (www.springer.com)

This book is dedicated to the valuable contribution made by all researchers in the field of Artificial Intelligence. Society continues to benefit from their hard work and inspiration. The editors also extend their unwavering support to those now involved in the sister domains of Computation Intelligence and Machine Intelligence. There is hope that these innovations will stimulate increased autonomy within the digital realm

Foreword

In the domain of Artificial Intelligence (AI) we are constantly challenged in investigating and modelling complex real-world problems and developing intelligent solutions. The field is rapidly being enriched by increases in processing power, communication infrastructure and distributed paradigms. The modern society is becoming increasingly savvy in the use of smart technologies in their business, social and professional lives. There is significant reliance on technology in the way we live our lives and in hybrid solutions that holistically solve real-world problems. This has reinvigorated enormous interest in research and innovation centred on AI. The new perspectives and interest have generated an exciting, yet active research interest that now underpins new applications.

Research in most areas of complex systems (well-structured or ill-structured) fit the characteristics of AI. The AI domain is diverse and many sub-domains form the current research frontier. Examples include: signal processing, knowledge representation (of structured and unstructured data), computational modelling, pattern recognition, data analytics, big data, behavioural modelling and knowledge discovery. They represent rich domains of research and innovation in the context of current and emerging needs of the industry and society. The modern paradigm of distributed computing is connecting computers, systems, organisations, individuals and communities. This adds new challenges for AI and empowers intelligent solutions. The potential of AI can be enormous but its full breadth is difficult to fully grasp. AI has produced some extraordinary success stories, however, it seems that the golden age of AI is just beginning. The renewed interest in this modern era of AI is primarily due to the growing technical capacity of researchers in solving more complex real-world problems being championed through industry support.

Jeffrey Tweedale and Lakhmi Jain have been leading contributors to AI research. They compiled the compendium of recent contributions to research and innovation in AI based on industry and real-world applications. They are both well-accomplished researchers in core areas of AI and problem-based applications. The perspective adopted in this compendium is of holistic research and innovation to model difficult real-world problems. The chapters selected for the book update on the modern perspective of AI research and innovation. It is a very good collection covering several areas of applications and motivates new research directions. The theme across all chapters combines several domains of AI research, especially computational intelligence and machine intelligence.

The first chapter introduces the recent research and models. Each of the subsequent chapters reveals leading edge research and innovative solution that employ AI techniques with an applied perspective. The problems include classification of spatial images, early smoke detection in outdoor space from video images, emergent segmentation from image analysis, intensity modification in images, multi-agent modelling and analysis of stress. They all are novel pieces of work and demonstrate how AI research contributes to solutions for difficult real-world problems that benefit the research community, industry and society. The book is informative for researchers and AI practitioners. It provokes ideas for further research directions. I have found it stimulating, and believe you will too.

Canberra, April 2013 Dharmendra Sharma

Preface

Artificial Intelligence (AI) is a field of study related to getting computers to achieve intelligent tasks. Research is predominantly software based, however, requires and incorporates technology to build new or innovative tools that ultimately benefit society. Industry continues to invest in AI, with many creating internal research facilities .The entertainment industry was a heavy investor in the twentieth century, although value engineers continue to mechanize industry, resulting in a shift towards more intelligent robots and unmanned systems. Another growth area is in the field of medical imaging, monitoring and portable healthcare products.

Two major research threads have manifested in semantic programming. These include: AI and Knowledge Engineering support of domain specific applications. Intelligent Systems are becoming ubiquitous in a wide range of situations. These include facets of simple everyday actions on mobile devices through to more advanced enterprise level applications in logistic systems and the medical domain. Society benefits daily, using applications to deliver digital news, socialisation and enhancements derived from expert decision making in knowledge-based systems. Knowledge Engineering relies on the exploitation of AI techniques to employ human-like intelligence in machine systems, tailored to solve specific problems. Hence, researchers continue to improve existing techniques within the domain.

This evolution in Information Processing has become a pervasive phenomenon within the community. Mobile computing continues to promote the ubiquitous access to information resources. Technology is enabling increased processing capabilities to hand-held devices, forcing more innovative access techniques to existing intelligent systems. Society is beginning to demand everyday applications that provide convenient access to the wealth of information processing systems serving the public. To achieve this, we must take advantage of the most recent research in information technologies.

We have chosen a dozen world class contributions from leading-edge researchers to provide readers with the ability to explore cutting edge examples of this evolution in a single volume. With over 50 years of combined experience in promoting and sharing advancements in AI and Knowledge-Based engineering, the editors are proud to offer this ensemble of experts and the wealth of knowledge presented. The reader is encouraged to read the introduction for initial direction prior to focusing on their topic of interest. They are also encouraged to extend their interest to the remaining chapters for an up-to-date exposure to a diverse range of modern AI topics and techniques. Several chapters are also dedicated to employing AI methodology to a diverse range of problems in the industry. We hope you enjoy the innovations presented as much as we did shepherding these contributions into print.

May 2013

Jeffrey W. Tweedale
Lakhmi C. Jain

Acknowledgments

This publication would not have been possible without the dedication and innovation of researchers within the field of Artificial Intelligence. The continuous evolution of techniques and methodologies is rapidly being applied to more diverse problems, ultimately seeding new domains. This culmination of multi-disciplined skills is being employed in the field of Computational Intelligence and continues to focus on enhancing Machine Intelligence. Obviously without the contribution of McCarthy, Minsky, Lippman, Zadeh and many others, existing researchers could be employed in completely different disciplines.

Similarly, without the efforts of the chairs who organised the 16th annual conference on Knowledge-Based and Intelligent Information and Engineering Systems[1], the ideas for this publication would not exist. The scene for this topic was stimulated through the rich source of contributions to this annual event. The authors were selected based on their initial contributions published in those proceedings as 'Advances in Knowledge-Based and Intelligent information and Engineering Systems 2012' [1]. These include topics relating to:

- Hyperspectral Imaging [2];
- Smoke Detection [3];
- Medical Imaging [4];
- Obtaining Shape from SEM Images [5];
- Position Fixing [6];
- Image Segmentation [7];
- Topology Migration [8];
- Vehicle Routing Problem [9];
- Body Condition Scoring [10];
- Web Performance Forecasting [11]; and
- Multi-Agent System Stress Café [12].

[1] Held in San Sebastian, Spain from 10 to 12 September 2012.

References

[1] Grãna, M., Toro, C., Posada, J., Howlett, R.J. Jain, L.C.: Advances in knowledge-based and Intelligent Information and Engineering Systems. In: 16th Annual KES Conference, San Sebastian, Spain, 10–12 September 2012. Frontiers in Artificial Intelligence and Applications, Vol. 243, IOS Press (2012)
[2] Quesada-Barriuso, P., Argüello, F., and Heras, D.B.: Efficient segmentation of hyperspectral images on commodity gpus. In: Grãna, M., Toro, C., Posada, J., Howlett, R.J. and Jain, L.C. (eds.) KES, Vol. 243 of Frontiers in Artificial Intelligence and Applications, pp. 2130–2139. IOS Press (2012)
[3] Favorskaya, M.N., Levtin, K.: Early smoke detection in outdoor space by spatio-temporal clustering using a single video camera. In: Grãna, M., Toro, C., Posada, J., Howlett, R.J. and Jain, L.C. (eds) KES. Vol. 243 of Frontiers in Artificial Intelligence and Applications., IOS Press pp. 1283–1292 (2012)
[4] Sierra, C.V., Novo, J., Reyes, J.S., Penedo, M.G.: Evolved artificial neural networks for controlling topological active nets deformation and for medical image segmentation. In Grãna, M., Toro, C., Posada, J., Howlett, R.J. and Jain, L.C. (eds.) KES. Vol. 243 of Frontiers in Artificial Intelligence and Applications. IOS Press pp. 1380–1389 (2012)
[5] Iwahori, Y., Shibata, K., Kawanaka, H., Funahashi, K., Woodham, R.J. and Adachi, Y.: Obtaining shape from sem image using intensity modification via neural network. In: Grãna, M., Toro, C., Posada, J., Howlett, R.J. and Jain, L.C. (eds.) KES. Vol. 243 of Frontiers in Artificial Intelligence and Applications. IOS Press pp. 1778–1787 (2012)
[6] Filipowicz, W.: Fuzzy evidence reasoning and position fixing. In: Grãna, M., Toro, C., Posada, J., Howlett, R.J., Jain, L.C., (eds.) KES. Vol. 243 of Frontiers in Artificial Intelligence and Applications. IOS Press pp. 1181–1190 (2012)
[7] Moreno, R., D'Anjou, A.: Hyperspectral image segmentation by t-watershed and hyperspherical coordinates. In: Grãna, M., Toro, C., Posada, J., Howlett, R.J., Jain, L.C., (eds.) KES. Vol. 243 of Frontiers in Artificial Intelligence and Applications., IOS Press pp. 2114–2121 (2012)
[8] Jedrzejowicz, P., Wierzbowska, I.: Impact of migration topologies on performance of teams of a-teams. In: Grãna, M., Toro, C., Posada, J., Howlett, R.J., Jain, L.C., (eds.) KES. Vol 243 of Frontiers in Artificial Intelligence and Applications., IOS Press pp. 1161–1170 (2012)
[9] Barbucha, D.: An agent-based implementation of the multiple neighborhood search for the capacitated vehicle routing problem. In: Grãna, M., Toro, C., Posada, J., Howlett, R.J., Jain, L.C., (eds.) KES. Vol. 243 of Frontiers in Artificial Intelligence and Applications., IOS Press pp. 1191–1200 (2012)
[10] Tedin, R., Becerra, J.A., Duro, R.J., Lede, I.M.: Towards automatic estimation of the body condition score of dairy cattle using hand-held images and active shape models. In: Grãna, M., Toro, C., Posada, J., Howlett, R.J., Jain, L.C., (eds.) KES. Vol. 243 of Frontiers in Artificial Intelligence and Applications., IOS Press pp. 2150–2159 (2012)
[11] Borzemski, L., Kaminska-Chuchmala, A.: Knowledge engineering relating to spatial web performance forecasting with sequential gaussian simulation method. In: Grãna, M., Toro, C., Posada, J., Howlett, R.J., Jain, L.C., (eds.) KES. Vol. 243 of Frontiers in Artificial Intelligence and Applications., IOS Press pp. 1439–1448 (2012)
[12] Ghosh, A., Tweedale, J.W., Nafalski, A., Dollard, M.: Multi-agent based system for analysing stress using the stress caf′e. In: Grãna, M., Toro, C., Posada, J., Howlett, R.J., Jain, L.C., (eds.) KES. Vol. 243 of Frontiers in Artificial Intelligence and Applications., IOS Press pp. 1656–1665 (2012)

Contents

Acronyms

AI	Artificial Intelligence
A-Team	Asynchronous Team of Agents
AA	Average Accuracy
ABiCA	Automatic Body Condition Assessment
ACT	Australian Capital Territory
AI	Artificial Intelligence
ALNS	Adaptive Large Neighborhood Search
ANN	Artificial Neural Network
API	Application Programming Interface
ASM	Active Shape Model
ASN	Autonomous Systems Number
AS	Autonomous Systems
AVIRIS	Airbone Visible-Infrared Imaging Spectrometer
AWB	Australian Workplace Barometer
BCS	Body Condition Score
BoW	Bag-of-Word
BRDF	Bidirectional Reflection Distribution Function
CA–Watershed	Watershed Transform based on Cellular Automata
CA	Cellular Automata
CBCT	Cone Beam Computed Tomography
CETA-CIEMAT	Centro de Extremadura Investigación de Tecnologías Avanzadas
CIDR	Classless Inter-Domain Routing
CI	Computation Intelligence
CMG	Color Morphological Gradient
CMYK	Cyan, Magenta, Yellow, BlacK
CS	Class Accuracy
CT	Computed Tomography
CUBLAS	CUDA Basic Linear Algebra Subroutines
CUDA	Compute Unified Device Architecture
CVRP	Capacitated Vehicle Routing Problem
DA	Decision Analyzer
DDR3	Double Data Rate type Three
DE	Differential Evolution
DFCLT	Dominant Flame Color Lookup Table

DRAM	Dynamic Random Access Memory
DRM	Dichromatic Reflection Model
DSS	Decision Support System
EA	Evolutionary Algorithm
EC	Evolutionary Computing
EPTSP	Euclidean Planar Traveling Salesman Problem
ERDF	European Regional Development Fund
ES	Evolutionary Strategies
FIPA	Foundation of Intelligent Physical Agents
FIS	Fuzzy Inference System
FIXMTE	Position Fixing with Mathematical Theory of Evidence
FLC	Fuzzy Logic Control
FMM	Fast Marching Method
FOPL	First Order Predicate Logic
FS-NEAT	Feature Selective Neuro-Evolution of Augmenting Topologies
FSM	Finite State Machine
FuSM	Fuzzy State Machines
GA	Genetic Algorithm
GDDR5	Graphics Double Data Rate type Five
GOFAI	Good Old-Fashioned Artificial Intelligence
GPGPU	General-Purpose Computing on Graphics Processing Units
GPU	Graphics Processing Units
GPU	Graphics Processing Unit
GP	Genetic Programming
GUI	Graphical User Interface
HCI	Human Computer Interaction
HDD	Hard Disk Drive
HF-NN	Hopfield like Neural Network
HSB	Hue, Saturation, Brightness
HSV	Hue, Saturation, Value
HTML	Hypertext Markup Language
HTTP	HyperText Transfer Protocol
HVRP	Heterogeneous Fleet Vehicle Routing Problem
IAF	Intelligent Agent Framework
IA	Intelligent Agent
IMADA	Intelligent Multi-Agent Decision Analyser
IoT	Internet of Things
IQ	Intelligence Quotient
JABATJ	JADE-Based A-Team Environment
JADE	Java Agent Development Framework
JSPS	Japan Society for the Promotion of Science
KB	Knowledge Base
KES	Knowledge-Based and Intelligent Engineering Systems
KIF	Knowledge Interchange Format
LBPV	Local Binary Pattern Variance

LBP	Local Binary Pattern
LC	Learning Component
LDA	Latent Dirichlet Allocation
LNS	Large Neighborhood Search
LS	Local Search
MAE	Mean Absolute Error
MAS	Multi-Agent System
MIMD	Multiple Instruction Multiple Data
MI	Machine Intelligence
MLP	Multi-Layered Perceptron
ML	Machine Learning
MOEA	Multiobjective Optimization Evolutionary Algorithm
MP	Minimum or Plateau
MQ	Machine Quotient
MRE	Mean Relative Error
MR	Magnetic Resonance
MTE	Mathematical Theory of Evidence
MV	Majority Vote
MWING	Multi-Agent Web pING
NEAT	Neuro-Evolution of Augmenting Topologies
NE	Neuro-Evolution
NM	Non-Minimum
NN	Neural Network
NSERC	Natural Sciences and Engineering Research Council
NSW	New South Wales
NT	Northern Territory
OAA	One-Against-All
OAO	One-Against-One
OA	Overall Accuracy
OD	Optic Disc
PDM	Point Distribution Model
PDVRPTW	Pickup and Delivery Vehicle Routing Problem with Time Windows
PDVRP	Pickup and Delivery Vehicle Routing Problem
PHP	Hypertext Preprocessor
pLSA	Probabilistic Latent Semantic Analysis
RBF-NN	Radial Basis Function Neural Network
RBF	Radial Basis Function
RCMG	Robust Color Morphological Gradient
RFID	Radio-frequency identification
RGB	Red, Green, Blue
RL	Reinforcement Learning
ROI	Region-Of-Interest
ROSIS	Reflective Optics System Imaging Spectrometer
RSK	Rules, Skills, Knowledge

rtNEAT	Real-time Neuro-Evolution of Augmenting Topologies
RTS	Real-Time Strategy
RTT	Round-Trip Time
SA	South Australia
SEM	Scanning Electron Microscope
SGS	Sequential Gaussian Simulation
SIMD	Single Instruction Multiple Data
SK	Simple Kriging
SLP	Single-Layer Perceptron
SM	Streaming Multiprocessor
SPEA2	Strength Pareto Evolutionary Algorithm 2
SQL	Structured Query Language
SVM	Support Vector Machine
TA-Teams	Team of A-Teams
TAN	Topological Active Net
TAS	Tasmania
TAV	Topological Active Volume
TB	Turning Bands
TSPLIB	Traveling Salesman Problem Library
TSP	Traveling Salesman Problem
UML	Unified Modeling Language
VDNS	Variable-Depth Neighborhood Search
VLNS	Very Large Scale Neighborhood Search
VNS	Variable Neighborhood Search
VRPTW	Vehicle Routing Problem with Time Windows
VRP	Vehicle Routing Problem
WA	Western Australia
WoT	Web of Things

Contributors

Yoshinori Adachi College of Business Administration and Information Science, Chubu University, 1200 Matsumoto-cho, Kasugai 487-8501, Japan

Francisco Argüello Centro de Investigación en Tecnoloxías da Información, University of Santiago de Compostela, Santiago de Compostela 15842, Rúa de Jenaro de la Fuente Domínguez, Spain

Dariusz Barbucha Department of Information Systems, Gdynia Maritime University, Morska 83, 81-225 Gdynia, Poland

J. A. Becerra Integrated Group for Engineering Research, University of A Coruña, Ferrol 15403, Spain

Richard J. Duro Integrated Group for Engineering Research, University of A Coruña, Ferrol 15403, Spain

Margarita Favorskay Siberian State Aerospace University, Krasnoyarsk 660014, Russia

Włodzimierz Filipowicz Faculty of Navigation, Gdynia Maritime University, Gdynia, Poland

Kenji Funahashi Department of Computer Science, Nagoya Institute of Technology, Gokiso-cho, Nagoya, Showa-ku 466-8555, Japan

Anusua Ghosh School of Electrical and Information Engineering, University of South Australia, Adelaide, Australia

Dora B. Heras Centro de Investigación en Tecnoloxías da Información, University of Santiago de Compostela, Santiago de Compostela, Rúa de Jenaro de la Fuente Domínguez, 15842, Spain

Yuji Iwahori Department of Computer Science, Chubu University, 1200 Matsumoto-cho, Kasugai 487-8501, Japan

Lakhmi C. Jain School of Electrical and Information Engineering, University of South Australia, Adelaide, Australia

Piotr Jędrzejowicz Department of Information Systems, Gdynia Maritime University, Morska 83, 81-225 Gdynia, Poland

Haruki Kawanaka School of Information Science and Technology, Aichi Prefectural University, 1522-3 Ibaragabasama, Nagakute 480-1198, Japan

Konstantin Levtin Siberian State Aerospace University, Krasnoyarsk, Russia

Ramn Moreno Computational Intelligence Group, Universidad del Pas Vasco, Donostia, Spain

Andrew Nafalski School of Electrical and Information Engineering, University of South Australia, Adelaide, SA, Australia

Jorge Novo Department of Computer Science, University of A Coruña, A Coruña, Spain

Manuel G. Penedo Department of Computer Science, University of A Coruña, A Coruña, Spain

Pablo Quesada-Barriuso Centro de Investigación en Tecnoloxías da Información, University of Santiago de Compostela, Rúa de Jenaro de la Fuente Domínguez, 15842 Santiago de Compostela, Spain

José Santos Department of Computer Science, University of A Coruña, A Coruña, Spain

Kazuhiro Shibata Department of Computer Science, Chubu University, 1200 Matsumoto-cho, Kasugai 487-8501, Japan

Cristina V. Sierra Department of Computer Science, University of A Coruña, A Coruña, Spain

R. Tedín Integrated Group for Engineering Research, University of A Coruña, Ferrol, Spain

Jeffrey W. Tweedale School of Electrical and Information Engineering, University of South Australia, Adelaide, Australia;

Air Operations Division, Defence Science and Technology Organization, Adelaide, Edinburgh, Australia

Izabela Wierzbowska Department of Information Systems, Gdynia Maritime University, Morska 83, 81-225 Gdynia, Poland

Robert J. Woodham Department of Computer Science, University of British Columbia, V6T 1Z4, Vancouver, B. C. Canada

Chapter 1
Advances in Modern Artificial Intelligence

Jeffrey W. Tweedale and Lakhmi C. Jain

Abstract This chapter presents a brief overview of advances in modern artificial intelligence. It recognises that society has embraced Artificial Intelligence (AI), even if it is embedded within many of the consumer products being marketed. The reality is that society is already in the throws of digitizing its past and continues progressively moves on-line. The volume and breadth of data being processed is becoming unfathomable. This digital future heralds the dawn of virtual communities, operating a Web of Things (WoT) full of connected devices, many fitted with wireless connectivity 24/7. This pervasiveness increases the demand on researchers to provide more intelligent tools, capable of assisting humans in prosecuting this information, seamlessly, efficiently and immediately. Ultimately AI techniques have been evolving since the 1950s. This evolution began with Good Old-Fashioned Artificial Intelligence (GOFAI) using explicitly coded knowledge, heuristics and axiomatization. This digital analogy of biological systems initially failed to realise its potential, at least until the birth of personal computers. This introduced a paradigm shift towards the Fuzzy/Neural era, which furnished society with computer vision, character recognition and Evolutionary Computing (EC) (among other successes). The value engineering proponents continued to invest in automation, which spurred the growth of Machine Intelligence (MI) research, further increasing expectations for computers to do more with less human interaction. McCarthy recently agreed

J. W. Tweedale (✉)
School of Electrical and Information Engineering, University of South Australia, Adelaide, Australia
e-mail: jeffrey.tweedale@unisa.edu.au; jeffrey.tweedale@dsto.defence.gov.au

L. C. Jain
Faculty of Education, Science, Technology and Mathematics, University of Canberra, Belconnen, Australia
e-mail: lakhmi.jain@unisa.edu.au

J. W. Tweedale
Aerospace Division, Defence Science and Technology Organization, Edinburgh, Adelaide, Australia

J. W. Tweedale and L. C. Jain (eds.), *Recent Advances in Knowledge-based Paradigms and Applications*, Advances in Intelligent Systems and Computing 234,
DOI: 10.1007/978-3-319-01649-8_1,

that it is now more appropriate to reliable AI research as Computational Intelligence (CI), because primitive methodologies have matured and science continues to witness more hybrid solutions. It is true that modern AI techniques typically employ multiple techniques and many now form hybrid systems with flexible problem solving capabilities or increased autonomy. This book contains a series of topics aimed at illustrating advances in modern AI. This book provides discussion on a number of recent innovations that include: classifiers, neural networks, fuzzy logic, Multi-Agent Systems (MASs) and several example applications.

Keywords Artificial Intelligence · Computational Intelligence · Evolutionary Computing · Fuzzy Logic · Machine Intelligence · Neural Network

1.1 Introduction

Information processing plays an important role in virtually all systems. This book examines a range of systems, that cover healthcare, engineering, aviation and education. This chapter presents some of the most recent advances in information processing technologies. A brief outline is presented with background information about knowledge representation and AI. A brief outline of each chapter is also included.

This book is intended to extend the readers knowledge of information processing and take you on a journey into many of the advanced paradigms currently experienced in this domain. There are as many forms of information as there are methods of prosecuting its sources. To achieve this goal researchers are required to communicate a collection of acquired facts, goals or circumstances and coalesce these repositories into one or more manageable bodies of knowledge. Society has increasingly become reliant on specialists to prosecute data reliably in order to make decisions about almost everything humans do.

Data is the representation of anything that can be meaningfully quantized or represented in digital form as a number, symbol and even text. Researchers process data into information by initially combining a collection of artefacts that are input into a system which is generally stored, filtered and/or classified prior to being translated into a useful form for dissemination. The processes used to achieve this task have evolved over many years and has been applied to many situations using a magnitude of techniques. Accounting and pay role applications take center place in the evolution of information processing. Data mining, expert system and knowledge-based system quickly followed. Today society lives in the information age in which people collect data faster than it can be processed. This book examines many recent advances in digital information processing with paradigms for acquisition, retrieval, aggregation, search, estimation and presentation.

Technically researchers could quote the abacus as being the first device used to process information. More recently the calculator, word processors and computing devices have had a major effects on society. Certainly the Internet became the single most disruptive influence in the modern era. It has provided access to information globally which is doubling exponentially. Societies ability to cope with this torrent of information continues to provide many challenges. Given that technology continues to provide improved access to even more sources of reliable data and faster machines to process information.

1.2 Artificial Intelligence

It is acknowledged that McCarthy first used the term Artificial Intelligence (AI) during a conference held in 1956 at *Dartmouth* [1]. Minsky prefers the term Computational Intelligence (CI)[1] [2], because it describes the science and engineering required to make machines conduct human-like task [3]. Others prefer to use the term Machine Intelligence (MI), however it is clear that scientific techniques continue to manifest as solutions to many industrial problems.

AI has endured a rocky history[2] for many reasons. Pollack believes that success is masked by the fact that solutions are embedded within systems as soon as they appear [4]. Progress is reported based on at least three phases of active research. These include Good Old-Fashioned Artificial Intelligence (GOFAI), an evolutionary period and modern AI techniques. Going forward a new wave of AI is forcast based on hybrid techniques using new cyber soultions. As industry continues to mechanize its workforce, it relies on CI methods to provide innovative solutions to provide increasingly more intelligent functionality [5–7]. Machines are starting to appear with human-like behavior. There is even an expectations that they will soon evolve limited congnitive functionality. Regardless of the terminlogy used, AI techniques continue to attract significant research effort because they do benefit society.

GOFAI research techniques used basic symbolic computation and is commonly referred to as Weak AI [8]. AI uses a collection of facts and heuristic knowledge to solve deterministic problems. Bourg and Seeman discussed a broader interpretation for its use in games [9] and a number of techniques have evovled to model the cognitive aspects of human behavior. Other developments include: perceiving, reasoning, communicating, planning and learning.

Alan Turing introduced a test to measure computation intelligence using two approaches (commonly known as top-down and bottom-up) [10, 11]. Most researchers initially followed formal programming techniques, using top-down symbolic AI approaches. Unfortunately cognitive functions are described using high-level concepts, requiring top-down techniques [12]. Researchers using the bottom-up approach produced simple elements that emulated primitve cognitive processes called Artificial Neural Networks (ANN). Following a series of leaps and bounds, researchers [13–15] realized that problems could be solved by incorporating learning method within Multi-layered Perceptron (MLP) using backpropagation. These developments stimulated a significant resurgence of ANN research [16–20].

[1] McCarthy also recently stated it would be more appropriate to use the term CI.

[2] AI is commonly being maligned based on expectation promoted through science-fiction movies.

There is a clear transition period where more Modern AI techniques appeared. Although no boundaries have been defined, many believe this transition is associated with the appearance of: Rule-Based systems [21, 22], Finite State Machines (FSM) [23], Fuzzy Logic [24, 25] and even Fuzzy State Machines (FuSM) [26].

Following another dip in research (introducing a possible advanced phase of AI), Russell and Norvig redefined AI as the 'study of creating systems that think or act like humans', alternatively it can be described as a rational approach to doing the 'right thing', within a known environment [7]. This new AI embodied rationality inside agents that receive inputs from the environment via sensors and provide outputs using effectors respectively. This definition has since been adopted by many in the AI community.

The concept of Multi-Agent Systems (MASs) emerged to tie together the isolated sub-fields of AI. A MAS consists of teams of Intelligent Agent (IA) that are able to perceive the environment using their sensory information, process the information with different AI techniques to reason and plan their actions in order to achieve certain goals [27, 28]. IA may be equipped with different capabilities including learning and reasoning. They are able to communicate and interact with each other to share their knowledge and skill to solve problems as a team. MASs have been used to create intelligent systems and they have a very promising future.

Advanced AI techniques include non-deterministic techniques that enable entities to evolve and learn or adapt [9]. Advanced ANN, Bayesian Networks, Evolutionary Algorithm (EA) and Reinforcement Learning (RL) have become mainstream preprocessors used in hybridised techniques. For instance:

- Bayesian Networks are used to enable reasoning during uncertainty [29].
- ANNs provide a relevant computational model used by agents to adapt to changes in the environment using learning techniques (either *Supervised, Unsupervised* or *Reinforcement*) [7]. Some examples of learning paradigms include: Temporal Difference learning [30] and Q-Learning [31].
- EA techniques are within the category of *Evolutionary Computation* and have been used for learning [32].
- Genetic Algorithm (GA), Genetic Programming (GP) [33], Evolutionary Strategies (ES) [34] and Neuro-Evolution (NE) [35]. GA techniques [36] also offer opportunities for optimise or evolve intelligent game behavior.
- NE [35] is a machine learning technique that uses EAs to train ANN. Examples of NE techniques include Neuro-Evolution of Augmenting Topologies (NEAT), Feature Selective NeuroEvolution of Augmenting Topologies (FS-NEAT) [37] and Real-time Neuro-Evolution of Augmenting Topologies (rtNEAT) [38, 39].

1.3 Research Directions

Many of the advanced techniques use Hybrid AI systems. Here, traditional AI techniques are used to pre-process uncertainty prior to using advanced AI techniques to solve real-world problems. The implementation of advanced AI techniques has provided researchers with many challenges because they are extremely difficult to understand, develop and debug. The lack of advanced AI technique experience by game developer has created a barrier to the expansion of these techniques in commercial games. The aim of this research is to provide appropriate tools in a test-bed to enable researchers investigate all forms of advanced AI techniques suited to problem solving.

Humans increasingly rely on machines to assist them in manufacturing and their everyday live. During the post war era, industry has embraced robotics and continues to mechanize its workforce. The trend began with logistic systems (packing, stacking, canning and bottling) and has steadily moved towards semi-autonomous articulated machines with nine degrees of freedom and these machines are no-longer fixed to static locations (painting, welding and assembly). Scientists continue to categorize robots based on their intended function [40]. A summary of several categories include:

Intelligent Tools:	Assistants to improve quality, efficiency and ultimately improve productivity;
Augmentation Systems:	Exo-skeletons for strength, agility and endurance; and
Environmental Assistants:	To protect humans from hazards, repetition and steady them during micro-manipulation.

General Motors Holden installed the *first* industrial robot (Unimate) in its New Jersey factory in 1961. This mechanized work force continues to evolve. Today there are between 50–250 robots per 10,000 employees[3] within the industrial nationals [41]. According to the International Federation of Robotics[4], this figure will grow by 200,000 installations per annum. China, India, Brazil and eastern Europe have clearly indentified the value chain propositions associated with increased scale and productivity tools. They have all forecast increased investment in future robotic manufacturing. Unfortunatley present day machines are still considered primitive with respect to human level intelligence, slowing their ability to displace humans engaged in cognitive activities. In fact there is no standard measurement for machine intelligence (sudo Machine Quotient (MQ)) and non of these measures have any relationship to the human Intelligence Quotient (IQ) rating or scale. It may not be necessary to proivde all machines with human-like intelligence, however it does make cognitive processing and self-governance easier to implement (representing the definition of true autonomy). At present, it is clear that increased automation insulates select industries from labour shortages and demographic shifts. It also improves long-term sustainability, eco-friendly production and power efficient measures to enable manufactures to increase their profit margins (globally). It should be possible for researchers to collaboratively employ 'intellignet capabilities' with highly autonomated sub-systems to improve the level of autonomy. Until then, more investment is required to realise more modern AI techniques to facilitate desired increases in

[3] In Australian terms this equates to only 0.5% of its total workforce.

[4] See www.worldrobotics.org.

levels of automation in the tools, assistants and intelligent systems being employed in the work place.

1.4 Contributions

All contributions in this book were sourced from authors attending the 16th annual conference on Knowledge-Based and Intelligent Information and Engineering Systems. A selection of topics and techniques have been selected to promote the innovation and advances in AI by active researchers in this field. The topics chosen reflect the need for designers to deliver smaller, faster, more portable computers. Industry is now manufacturing powerful Graphics Processing Units (GPUs) and smart phones. These devices have changed the way society accesses information and increases the demand for researchers to deliver increased auotomation through more innovative techniques to deliver desired outcomes to the public. Hence the editors focused on the selection focus on selection of techniques that contribute to this evolution. They include: classifiers in Sect. 1.4.1, neural networks in Sect. 1.4.2, fuzzy logic in Sect. 1.4.3, MASs in Sect. 1.4.4 and a number of applications used to solve industrial problems in Sect. 1.4.5. A full list of each chapter is show in Table 1.1.

Table 1.1 List of topics

Chapter	title
1	Advances in modern artificial intelligence
2	Computing efficiently spectral-spatial classification of hyperspectral images on commodity GPUs
3	Early smoke detection in outdoor space by spatio-temporal clustering using a single video camera
4	Using evolved artificial neural networks for providing an emergent segmentation with an active net model
5	Shape from SEM image using fast marching method and intensity modification by neural network
6	Fuzzy evidence reasoning and navigational position fixing
7	Segmentation of hyperspectral images by tuned chromatic watershed
8	Impact of migration topologies on performance of teams of agents
9	A cooperative agent-based multiple neighborhood search for the capacitated vehicle routing problem
10	Building the automatic body condition assessment system (ABiCA), an automatic body condition scoring system using active shape models and machine learning
11	The impact of network characteristics on the accuracy of spatial web performance forecasts
12	Using multi-agent systems techniques for developing an autonomous model used to analyze work-stress data

1.4.1 Classifiers

Most AI applications use classifiers to either divide (segregate data) or control the flow of information. They can also be used to pre-process data into categories, prior to inferring any further action(s). The most popular techniques include: random forest, Neural Networks (NNs), Support Vector Machines (SVMs), k-nearest neighbour, Bayes and decision trees. The system designer will select which technique is appropriate based on the desired performance, efficiency and the type of data being prosecuted [42]. It is not possible to describe every classifier in this volume, therefore two SVM examples have been provided. One using traditional techniques on a standard computer network for smoke detection and the other embracing GPU enhancements for hyperspectral imaging.

Hyper-spectral Imaging. Hyper-spectral imaging is the process of analysing high resolution, multi-spectrum images to measure properties previously unavailable using the visible light spectrum. Using new technology, sensors are now able to capture hyper-spectral images (visible, infra-red and ultraviolet light bands). The mining industry has invested in this technology, because every chemical has its own unique response to light at different wavelengths. A spread spectrum sensor combines the science of spectroscopy and imaging technologies in order to conduct for mineral exploration [43]. Most sensors provide a serial stream of data. The volume of data increases as the spectrum increases, hence the processing required must also increase. A supercomputer processes increased volume with parallel processes, typically using a Single Instruction, Multiple Data (SIMD) taxonomy [44]. Taking these machines into the field is as prohibative as maintaining an adhoc broad-band connection. GPUs are providing access to this form of technology to portable computers, facilitating this process on aerial platforms. It is possible to accelerate geometric calculations; such as the rotation and translation of vertices into different coordinate systems, using Compute Unified Device Architecture (CUDA) and Open Graphics Library (OpenGL) or more recently Open Computing Language[5] (OpenCL).

Recently pixel-wise segmentation based on SVM classifiers have emerged. Tarabalka et al. [45] incorporated a watershed transform technique to conduct classification of hyperspectral images. They obtained a Robust Color Morphological Gradient (RCMG) using readily available General-Purpose Computing on Graphics Processing Units (GPGPUs). Similarly Quesada-Barriuso et al. also introduced an innovative technique for segmentation and classification of hyper-spectral images using GPUs [46]. Their scheme efficiently exploits memory hierarchy and using thousands of threads facilitated within the GPU architecture. Chapter 2 documents experiments used to analyse five techniques (Two using SVM, one Robust Color Morphological Gradient (RCMG), Asynchronous CA-Watershed and the Majority vote algorithms). The results were impressive, with the asynchronous CA-Watershed achieving an increase of up to 27.5 times that of the baseline.

---

[5] This is a standard to facilitiate the development of code for both the CPU and GPUs with an emphasis on achieving accelerated computation throughput for portable processing.

Smoke Detection. Most people in society associated smoke as an environmental consequence of combustion. Scientist know it can occur as a aerosol or condensed phase components typically contain evolved gases. The appearance and structure, ranges from from light colored whiffs or clouds to turbulent sooted swells. These flow patterns will determine how the smoke concentrates and moves within its environment. The properties are therefore based on color, concentration and the distribution of particles (size, density and distribution). This makes smoke detection difficult and typically reliant on visual methods. Photoelectric and ionization sensors are currently used to detect smoke in many commercial, industrial and residential systems. Automated fire alarms originated in Boston during 1852 (Mosser and Farmer).[6] These indicated the location of a *fire*. They were aimed at reducing the loss of life and property.

Moving forward, *heat* detectors were patented by George Andrew Darby in 1902. Integrated fire alarm systems started appearing in the 1930's, however battery operated detectors flourished during the 1960's and since then, their mandatory installation has become law for many nations. Existing detectors rely on the detection of smoke, robustly and as early as possible. Verstockt et al. [47] evovled a smoke and flame detection systems using visual sensors. Researchers have embraced Computer vision using AI techniques to map and track smoke and flame, even at night. Toreyin et al. [48] incorporated techniques to detect motion, edge blurring, and color features within smoke wavelets. Stream dection and texture pattern analysis using hidden Markov trees were subsequently proposed by Ferrai et al. [49]. New techniques are emerging based on spatio-temporal clustering. Favorskaya and Levtin explain an advanced algorithm used to extract an enhanced set of features from video sequences in Chap. 3.

This describes the process of detecting smoke outdoors using Spatio-temporal Clustering techniques on images provided from a commercially available video camera [50]. It uses commercial Red, Green, Blue (RGB) video streams and their results are promising. This technology indicates they can provide an effective and inexpensive method of identifying smoke and fire alarm for use in the outdoor space using a real-time application. Color-texture analysis of moving regions includes the choice of the color space and application of filters. An approach based on moving clusters will be appropriate in detecting smoke dynamically. This investigation uses statistical self-similarity and a wavelet analysis to improve the techniques used to identify smoke segmentation. Although physical detection is possible, augmented visual systems provide improved early warning and reduced false alarms, especially in dangerous situations.

[6] See http://www.alliedfiresafety.com/page/alarms/.

1.4.2 Neural Nets

This topic of NNs focuses on artificial means of mimicking a biologically inspired nervous system. The key elements of an ANN are the perceptrons (neurons), interconnected processing elements (neurones), its structure and weights [51]. Lipmann provides a rounded description, reviewing six important neural net models [52]. These include the: classic M class classifiers [53], Hopfield nets [54], Hamming nets [55], Carpenter/Crossberg classifiers [56], Single-Layer Perceptrons (SLPs) [15] and Kohonen's self organising feature maps [57]. Typical applications include pattern recognition or data classification, however more novel applications continue to apear. Two examples include: medical image segmentation and shape formation using SEM images.
Medical Image Segmentation. Computer-aided diagnostic systems continue to attracting significant research effort. There are obvious cost benefits for the patient, their specialist and even the diagnostic provider. Numerous techniques have evolved to support segmentation. Some include: thresholding, clustering, compression, Histogram, Edge detection, region-growing, split-and-merge, partial differential equations, graph partitioning, watershed and model-based segmentation. Novo et al. describe a new approach to segmentation of medical images using Topological Active Net (TAN) deformable models [58]. They use a discrete implementation of an elastic $n-$dimensional mesh with interrelated nodes [59]. Their methodology was tested using Computed Tomography (CT) images to evolve an ANN with a Differential Evolution (DE) solution using between 100 and 150 images. Chapter 4 discusses their design and the test methodology. Validation was performed using CT, Cone Beam Computed Tomography (CBCT) and retinal images.
Obtaining Shape from SEM Images. A Scanning Electron Microscope (SEM) uses a focused electron beam to methodically illuminate the surface of its target material. Individual electrons interact with the material being bombarded and the detected signals used to generate the topological structure and composition. Iwahori et al. use a Fast Marching Method (FMM) to control the Intensity Modification process to generate an accurate 3D image [60]. The optimization is applied using the Hopfield like Neural Network (HF-NN) [61]. Where the initial vectors are determined by the Radial Basis Function Neural Network (RBF-NN), before being transferred to the HF-NN and optimized. The NN used to control the image intensity was trained with a variety of learning epochs (100, 200 and 300) in order to determine the optimal settings. These experiments are validated through comparision of samples images in Chap. 5.

1.4.3 Fuzzy Logic

Zadeh offered his 'fuzzy set theory' in 1965 [25]. He proposed the use of linguistic variables to manage groups of crisp sets used to represent infinite-valued logic [62]. This theory represented a revolutionary control theory within the field of AI. A degree of truth can now be represented using words that more accuratly approximated human concepts. For instance speed could be presented to humans within context as 'fast' or 'slow', rather than x or y. A new series of linguistic variables were introdcued to facilitate the expression of rules and facts. These would be interpreted within machines using inference systems, embedded within a Fuzzy Logic Control (FLC). Two common styles of FLC include Sugeno [63, 64][7] and Mamdani [65, 66]. Using a Sugeno Fuzzy Inference System (FIS), goal reaching functionality can be achieved by reviewing the error data at each range interval. For instance, the required distance measurements and bearing can be calculated using fuzzified variables and projecting crisp (de-fuzzified) coordinates.[8]

Using Fuzzy Logic, machines are able to cope with unknown, unstructured and dynamically changing environments. New systems evolved that are capable of fuzzy approximation, recognition and behavior modeling using human-like terminology.[9] Recent examples include improve position fixing while navigating and improved image segmentation while processing hyper-spectral scenes.

Position Fixing. By extending the concepts associated with membership, an extended fuzzy environment can be used to create new opportunities that enable problem solving with uncertainty. Many real-world issues contain uncertainty, for instance the art of navigation requires dead-reckoning against known artefacts, however when none are available, uncertainty becomes a factor. Technology helps to remove uncertainty, but many solutions still employ statistical methods to approximate the reported position. For instance, in radar systems, all radar bearings are resolved against a mean error within a specified range [67]. Models that include uncertainty can be created by applying Mathematical Theory of Evidence (MTE). This theory can be perceived of as an extension of the Bayesian Concept [68]. This technicque offers a mechanism to enrich information within the context of the initial evidence. In the example offered by Filipowicz [69], evidence is encoded into a belief structures prior to being combined into the final estimate [70].

In an extended version of this paper, measurements and indication data, along with nautical knowledge, was encoded into belief functions [71]. This knowledge and associated data are considered as evidence that is exploited while navigating. The Belief functions in their application represent evidence that is treated in order to increase its informative context. The representation of this knowledge and its generated results provide pseudo belief structures. In Chap. 6 they postulate that any conflict occurring within beliefs system provides greater plausibility, supporting the generation of "a navigational fix" (within their limited range of test cases).

Image Segmentation. Spread spectrum image sensors are increasingly being used to extend computer vision technology. This in-turn places pressure on researcher to deliver sensors that process light beyond the RGB spectrum. The terms multi-spectral image processing and hyper-spectral images have began flooding scientific litera-

---

[7] See also Tsukamoto from 1979 and later 'TSK' from 1985.

[8] Example variables include: X_i, Y_i, Z_i and X_p, Y_p, Z_p.

[9] Examples include a number of simple well defined behaviors; such as: avoidance, reach, follow, align, jump, turn and pass.

ture. Articles describing many new algorithms describe the science associated with transforming broad bandwidth images into information used by real-time industrial applications.

New techniques have emerged to segment image content into usable objects. The concept is straight forward, however the implementation is an intrinsically difficult problem. Moreno et al. describe a segmentation algorithm for hyper-spectral and multi-spectral images [72, 73]. They use hyperspheres to encapsulate sub-sections known as n-spheres. Given these hyper-specrtal images can contain more than 30 bands (dimensions), this process is called Hyperspherical Transformation. Moreno and Graña use a t-Watershed algorithm to process images provided via an SOC 710 camera with 128 bands [74]. They extend this concept to deliver an improved version of the watershed transformation in Chap. 7. This provides good segmentations and avoids well-known problems, like; the effect of changes on illumination and the standard oversegmentation.

1.4.4 Multi-agent Systems

The concept of MASs emerged to tie together the isolated sub-fields of AI. A MAS consists of teams of IA that are able to perceive the environment using their sensory information, process the information with different AI techniques to reason and plan their actions in order to achieve certain goals [27, 28]. IA may be equipped with different capabilities including learning and reasoning. They are able to communicate and interact with each other to share their knowledge and skill to solve problems as a team. MASs have been used to create intelligent systems and they have a very promising future. For further background, the reader can consult a survey announcing current innovations in multi-agent systems [75]. However for a more detailed description, access to advances in information processing paradigms [76] is recommended. This book describes MAS techniques associated with migrating the topology of agents and another solving the travelling salesman or vehicle routing problem.

Topology Migration. Local search has often been used to solve combinatorial optimization problems [77]. An evolution of these techniques has facilitated the development of parallel and distributed versions of population based methods. Wooldridge championed a new paradigm of population-based methods by embedding these algorithms within one or more agents [78]. MAS architectures flourished, for example: the introduction of (JESS) [79], JACK[10] and JADE. A recent applicaiton employing agency research has been expressed using Asynchronous Team of agents (A-Team) [80].

Jędrzejowicz and Wierzbowska recently employed JADE to generate a Foundation for Intelligent Physical Agents (FIPA)[11] team of A-Team called (JABAT). Chapter 8

[10] Agent Oriented Systems—subsumed by CAE International.

[11] See http://www.fipa.org/.

discusses the results of their experiments to verify that the choice of the migration topology and frequency of migration influence a systems performance. They progressively examine the results obtained using JABAT to validate this hypothesis.

Vehicle Routing Problem. Multiple neighborhood search is another technique embedded within the MAS paradigm. Heuristic-based can be guided to avoid the solution being trapped in a local optimum. Using a cooperative strategy that focuses on agents running in parallel (using divergent algorithms), multiple neighborhoods can be considered. Barbucha proposes to conduct a computational experiment to examine the multiple neighborhood using search parameters [81]. For example using simulated annealing [82] that employs a modified, penalized goal function to guide its Local Search [83]. This approach could apply smoothing [84] or exploit multiple neighborhoods using Variable Neighborhood Search [85], Very Large Scale Neighborhood Search [86], or Adaptive Large Neighborhood Search [87].

Chapter 9 explores the concept of using a MAS to solve the local search optimization problems. This novel approach, employes multiple neighborhood search capabilities with a cooperative management function to explore different heuristics and assured diversification of the search using a parallel computation. The experiments conducted focused on solving instances of the Capacitated Vehicle Routing Problem (CVRP). The results verify improvements by comparing a variety of successive neighborhood search algorithms with a series of divergent goals.

1.4.5 Applications

Modern AI applications typically focus on Machine Learning techniques to solve industrial problems. Many employ multiple CI techniques to provide constrained solutions to complex problems. These systems are progressively being aggregated to form hybrid solutions. Unfortunately the problem space for Real-world problems is associated with a vast array of parameters. Systems have evolved that harness sub-sets of dynamic situations, however many fail to manage in hostile environments. Researchers are increasingly adopting hybrid approaches to solve industrial problems (where the environment can be controlled). Several examples include: a body condition scoring system for livestock, a forecasting system to predict Web Performance and a Stress Café.

Body Condition Scoring. Technology has enabled researchers to provide automation for many human-like activities, especially those using machinery. As society digitises its future, they continue to connect distributed functionality once segregated by time and space. Dairy farmers are still engaged in manual activities conducted in a time poor environments. A number of technologies are steadily being introduced to assist with growing produce and increasing the levels of automation associated with animal husbandry. Part of any evolutionary manager is the selection and scoring mechanisms. At present selection is managed using a grading system, using factors like age and body condition. At present the latter is a subjective assessment that is routinely conducted by human experts. This is time consuming and influenced by geographical and environmental factors.

Tedín et al. [88] have decomposed the process in order to build an autonomous equivalent. Here they discuss the concepts associated with generating an automatic body condition scoring system for dairy cattle using images using a hand-held cameras. Firstly, the shape of a cow is found and then the body condition score is estimated using this shape. Automatic Body Condition Assessment (ABiCA) uses an Active Shape Model (ASM) [89], tuned with an evolutionary algorithm [90]. Chapter 10 discusses the process of feeding symbol shapes into a regression function to evolve an estimated body condition score for each target animal. The results are quite promising, although not conclusive. More effort is required to generate usable images from a hand held camera, in cluttered and poorly illuminated environments on the farm.

Web Performance Forecasting. The use of spatial forecasting in the computer science domain is still in its infancy. Geostatistical methods have been used for simulation and estimation of sporting and species activities. As the Web of Thingss (WoTs) [91, 92] continues to evolve under the influence of adhoc and Bring Your Own Devices (BYODs) measuing performance for Quality Of Service (QOS) becomes a challenge. To assist in this pursuite, Borzemski and Kamińska-Chuchmała developed a proposal to measure Internet Web Performance forecasting [93].

Chapter 11 presents an overview of using geostatistical methods to analysis the performance and impact of various performance parameters while forecasts the QOS for servers belonging to different Autonomous Systems (AS). The geostatistical methods originate from the Kriging - the estimation method developed by Daniel Krige [94]. In this chapter they describe forecasting in terms of the geostatistical methods employed. The data for this research was collected in active Internet Web Performance measurements carried out by software agents monitoring a group of Web servers. In this research, data was collected using Multi-agent Web pING (MWING) applications based on networks in Gdańsk and Wrocław (to/from European Web servers) [95, 96]. This research culminated in the collection of large-scale measurements from many sources over a variety of time-frames and expected workloads. The calculations derived from the measurements collected, do show there is relationship between forecasts accuracy and fluent traffic on routes to servers. It also confirms that geographical distance also effects the accuracy of each forecast, although the physical distance of the networks backbone made little difference.

MAS Stress Café. Agency techniques have been hailed as the revolutionary breakthrough in software engineering because they can be used to encapsulate one or more capability within a single applications. Ghosh et al. presents a novel adaptation of agent capabilities to develop an hybridized autonomous MAS to analyze work-related stress data [97]. The Intelligent Multi-Agent Decision Analyser (IMADA) is the core component of this model, and each agent capability provides an independent capability that operates specified functionality in their own right. This framework manages the interaction and communicate between all agents to accomplish the desired goal. This semi-autonomous model replaces a costly and cumbersome manual approach, where data collected relied on phone calls to select communities.

IMADA uses a semi-autonomous data collection and analysis process to facilitate measuring psychological stress in the workplace. It generates a Kiosk-style approach for providing on-line, remote access to professional skills through a distributed series of workplace sites and industry cross sections. This application provides heuristic computation, with ANN classifiers and Fuzzy Logic feedback systems. The ANNs are capable of transforming an input vector from n-dimensional space to an output vector in m-dimensional space [98]. A Fuzzy logic grading system is applied to interpret the output from a neural network, as well as giving human-readable interpretation of the data in linguistic terms [99]. This hybrid intelligence provides the computational framework that integrates these techniques to support problem solving and decision making [100, 101]. Chapter 12 discusses the evolution of the work stress café application. Future work may include an emotion recognition agent, enabling more in-depth analysis of work stress related data. A learning component would be useful to add functionality and enable the model to evolve its analysis in real time.

Readers are reminded that this book presents a series of world class contributions, from leading-edge researchers, covering an array of advanced AI techniques. The reader is initially encouraged to focus on their topic of interest and then stimulate their imagination by reading the remaining chapters. These chapters have been selected from a diverse range of modern AI topics and techniques, employing hybrid methodologies to solve specific problems for industry. We hope you enjoy the innovations presented.

References

1. McCarthy, J.: Programs with common sense. In: Symposium on Mechanization of Thought Processes, National Physical Laboratory, Teddington (1958)
2. McCorduck, P.: Machines Who Think, pp. 1–375. Freeman, San Francisco (1979)
3. Minsky, M.: Society of Mind. Simon and Schuster, Pymble, Australia (1985)
4. Baard, M.: Ai founder blasts modern research. Wired News, pp. 1–2 (2003)
5. Nilsson, N.: Artificial Intelligence: A New Synthesis. Morgan Kaufmann Publishers (1998)
6. Poole, D., Mackworth, A., Goebel, R.: Computational Intelligence: A Logical Approach. Oxford University Press, New York (1998)
7. Russell, S.J., Norvig, P.: Artificial Intelligence: A Modern Approach. 2nd edn. Prentice Hall, Pearson Education, Inc., Upper Saddle River (2003)
8. Haugeland, J.: Artificial Intelligence: The Very Idea. MIT Press, Cambridge (1985)
9. Bourg, D.M., Seeman, G.: AI for Game Developers. O'Reilly, Media (2004)
10. Turing, A.: Intelligent machinery. In: Meltzer, D. (ed.) Machine Intelligence. vol. 5, Orginally, A National Physics Laboratory Report, pp. 3–23. Edinburgh University Press, (1948)
11. Turing, A.: Computing machinery and intelligence. In: Mind. vol. 59(236). Unpublished until **1968**, 433–460 (1950)
12. Jones, M.T.: AI Application Programming. Charles River Media, Inc. Hingham (2003)
13. Ackley, H., Hinton, E., Sejnowski, J.: A learning algorithm for boltzmann machines. Cogn. Sci. **9**, 147–169 (1985)
14. Hopfield, J.: Neurons with graded responses have collective computational properties like those of two-state neurons. In. Proceedings of the National Academy of Sciences (USA), vol. 81. pp. 3088–3092 (1984)

15. Rumelhart, D., Hinton, G., Williams, R.: Learning internal representations by error propagation. In: Rumelhart, D., McClelland, J. (eds.) Parallel Distributed Processing: Explorations in the Microstructure of Cognition, vol. 1. MIT Press, Cambridge (1986)
16. Carpenter, G., Grossberg, S.: Art 2: self-organization of stable category recognition codes for analog input patterns. Appl. Opt. **26**(23), 4919–4930 (1987)
17. Grossberg, S.: Competitive learning: from finteractive activation to adaptive resonance. Cogn. Sci. **11**, 23–63 (1987)
18. Kirkpatrick, S., Gelatt, C.D., Vecchi, M.P.: Optimization by simulated annealing. Science **220**(4598), 671–680 (1983)
19. Linsker, R.: Self-organization in a perceptual network. Computer **21**(3), 105–117 (1988)
20. Sutton, R.S., Barto, A.G.: Reinforcement Learning: An Introduction. MIT Press, Cambridge (1998)
21. Feigenbaum, E., McCorduck, P., Nii, H.P.: The Rise of the Expert Company. Times Books, New York (1988)
22. Jackson, P.: Introduction to Expert Systems. 3rd edn. Addison-Wesley (1999)
23. Hennie, F.C.: Finite-State Models for Logical Machines. Wiley, New York (1968)
24. Ross, T.J.: Fuzzy Logic with Engineering Application, 3rd edn. Wiley, Chichester (2010)
25. Zadeh, L.A.: Fuzzy sets. Inform. Control **8**(3), 338–353 (1965)
26. Grantner, J., Patyra, M.: Synthesis and analysis of fuzzy logic finite state machine models. In: Fuzzy Systems, : World Congress on Computational Intelligence, vol. 1, pp. 205–210. IEEE Press, Piscataway (1994)
27. Jennings, N., Wooldridge, M.: Software agents. IEE Review, Institut. Eng. Technol. **42**(1), 17–20 (1996)
28. Wooldridge, M., Muller, J., Tambe, M.: Agent theories, architectures, and languages: a bibliography. In: Intelligent Agents II Agent Theories, Architectures, and Languages, pp. 408–31. Springer, Berlin (1996)
29. Mackworth, A.: The coevolution of AI and AAAI. AI Mag. **26**, 51–52 (2005)
30. Sutton, R.S.: Learning to predict by the methods of temporal differences. Mach. Learn. **3**, 9–44 (1988)
31. Watkins, C.J.C.H., Dayan, P.: Technical note: Q-learning. Mach. Learn. **8**(3), 279–292 (1992)
32. Hughes, E.: Checkers using a co-evolutionary on-line evolutionary algorithm. In: The 2005 IEEE Congress on Evolutionary Computation, 2005. vol. 2, pp. 1899–1905 (2005)
33. Koza, J.: Genetic Programming: On the Programming of Computers by Means of Natural Selection. MIT Press, Cambridge (1992)
34. Beyer, H.: The Theory of Evolutionary Strategies. Springer, Berlin (2001)
35. Nolfi, S., Elman, J.L., Parisi, D.: Learning and evolution in neural networks. Technical report, Technical Report 9019, Center for Research in Language, University of California, San Diego (1990)
36. Holland, J.: Adaptation in Natural and Artificial Systems: An Introductory Analysis with Applications to Biology. Control and Artificial Intelligence. MIT Press, Cambridge (1975)
37. Stanley, K.O., Miikkulainen, R.: Evolving neural networks through augmenting topologies. Technical Report AI2001-290, Department of Computer Sciences, The University of Texas at Austin (2002)
38. Stanley, K.O., Bryant, B.D., Miikkulainen, R.: Evolving neural network agents in the nero video game. In: Proceedings of the IEEE 2005 Symposium on Computational Intelligence and Games (CIG'05), IEEE, Piscataway (2005)
39. Stanley, K.O., Bryant, B.D., Miikkulainen, R.: Real-time neuroevolution in the nero video game. IEEE Trans. Evol. Comput. **9**, 653–668 (2005)
40. Elfes, A.: Why the australian manufacturing industry needs the next generation of robots. In: The Conversation, CSIRO, Canberra, pp. 1–4 (2013)
41. Johnson, G.: The advance of the robotis. Whats New Process Technol. **26**(9), 4–7 (2013)
42. Hand, D.J.: Measuring Classifier Performance: A Coherent Alternative to the Area under the Roc Curve, Machine Learning, vol. 77, pp. 103–123. Springer-Velag, Berlin (2009)

43. Berman, M., Kiiveri, H., Lagerstrom, R., Ernst, A., Dunne, R., Huntington, J.: Ice: a statistical approach to identifying endmembers. IEEE Trans. Geosci. Remote Sensing **42**, 2085–2095 (2004)
44. Chai, S.M., Antonio, G., Lugo-Beauchamp, W.E., Cruz-Rivera, J.L., Wills, D.S.: Hyperspectral image processing applications on the simd pixel processor for the digital battlefield. In: Computer Vision Beyond the Visible Spectrum: Methods and Applications (CVBVS '99), pp. 130–138. IEEE Press, Piscataway (1999)
45. Tarabalka, Y., Chanussot, J., Benediktsson, J.A.: Segmentation and classification of hyperspectral images using watershed transformation. Pattern Recogn. **43**(7), 2367–2379 (2010)
46. Quesada-Barriuso, P., Argüello, F., Heras, D.B.: Efficient segmentation of hyperspectral images on commodity gpus. In: Graña, M., Toro, C., Posada, J., Howlett, R.J., Jain, L.C., (eds.) KES. Frontiers in Artificial Intelligence and Applications, vol. 243, pp. 2130–2139. IOS Press (2012)
47. Verstockt, S., Merci, B., Lambert, P., van de Walle, R., Sette, B.: State of the art in vision-based and smoke detection. In: Proceedings of the 14th International Conference on Automatic Fire Detection, vol. 2, pp. 285–292 (2009)
48. Toreyin, B.U., Dedeoglu, A.Y., Cetin, E.: Wavelet based real-time smoke detection in video. In: Proceedings of the 13th European Signal Processing Conference EUSIPCO, pp. 4–8 (2005)
49. Ferrari, R.J., Zhang, H., Kube, C.R.: Real-time detection of steam in video images. Pattern Recogn. **40**(3), 1148–1159 (2007)
50. Favorskaya, M.N., Levtin, K.: Early smoke detection in outdoor space by spatio-temporal clustering using a single video camera. In: Graña, M., Toro, C., Posada, J., Howlett, R.J., Jain, L.C. (eds.): KES. Frontiers in Artificial Intelligence and Applications, vol. 243, pp. 1283–1292. IOS Press (2012)
51. McCulloch, W.S., Pitts, W.H.: A logical calculus of the ideas immanent in nervous activity. Bull. Math. Biophys. **5**, 115–133 (1943)
52. Lippmann, R.: An introduction to computing with neural nets. ASSP Magaz. IEEE **4**(2), 4–22 (1987)
53. Duda, R., Hart, P.E.: Pattern Classificafion and Scene Analysis. John Wiley and Sons, New York (1973)
54. Hopfield, J.J.: Neural networks and ptiysical systems with emergent collective computational abilities, pp. 2554–2558. National Academy of Science, NSF, Washinton, DC (1982)
55. Wallace, D.J.: Memory and learning in a class of neural models. In: Bunk, B., Mufter, K.H. (eds.) Workshop on Lattice Cauge Theory, Wuppertal, Plenum (1986)
56. Carpenter, G.A., Grossberg, S.: Neural dynamics of category learning and recognition: attention, memory consolidation, and amnesia. In: Davis, J., Newburgh, R., Wegman, E. (eds.) Brain Structure, Learning and Memory. AAAS Symposium Series (1986)
57. Kandel, E.R., Schwartz, J.H.: Principles of neural circuits: a model. Science **233**, 625–633 (1986)
58. Sierra, C.V., Novo, J., Reyes, J.S., Penedo, M.G.: Evolved artificial neural networks for controlling topological active nets deformation and for medical image segmentation. In: Graña, M., Toro, C., Posada, J., Howlett, R.J., Jain, L.C. (eds.) KES. Frontiers in Artificial Intelligence and Applications, vol. 243, pp. 1380–1389. IOS Press (2012)
59. Ansia, F., Penedo, M., Mariño, C., Mosquera, A.: A new approach to active nets. Pattern Recogn Image Anal **2**, 76–77 (1999)
60. Iwahori, Y., Shibata, K., Kawanaka, H., Funahashi, K., Woodham, R.J., Adachi, Y.: Obtaining shape from sem image using intensity modification via neural network. In: Graña, M., Toro, C., Posada, J., Howlett, R.J., Jain, L.C. (eds.) KES. Frontiers in Artificial Intelligence and Applications, vol. 243, pp. 1778–1787. IOS Press (2012)
61. Hopfield, J.J. and Tank, D.W.: "Neural" computation of decisions in optimization problems. Biol. Cybernet. **52**, 141–152 (1985)
62. Lukasiewicz, J.: The logic of trivalent. Mov. Philos. **5**, 169–171 (1920)
63. Sugeno, M. (ed.): Industrial applications of fuzzy control. Technology and, Engineering (1985)
64. Takagi, T., Sugeno, M.: Fuzzy identification of systems and its applications to modeling and control. In: IEEE Transaction of Systems, Man, and Cybernetics. vol. 15(1), pp. 116–132. IEEE Press, Piscataway (1985)

65. Mamdani, E.: Application of fuzzy algorithms for control of simple dynamic plant. Proc. Instit. Electr. Eng. **121**(12), 1585–1588 (1974)
66. Sivanandam, S., Sumathi, S., Deepa, S.: Introduction to fuzzy logic using MATLAB. Springer, New York, NY (2007)
67. Jurdziński, M.: Principles of Marine Navigation. WAM, Gdynia (2008)
68. Staker, R.: Use of bayesian belief networks in the analysis of information system network risk. Information, Decision and Control, IDC 99. Proceedings. pp. 145–150 (1999)
69. Filipowicz, W.: Fuzzy evidence reasoning and position fixing. In: Graña, M., Toro, C., Posada, J., Howlett, R.J., Jain, L.C., (eds.): KES. Frontiers in Artificial Intelligence and Applications, vol. 243, pp. 1181–1190. IOS Press (2012)
70. Denoeux, T.: Modelling vague beliefs using fuzzy valued belief structures. Fuzzy Sets and Syst. **116**, 167–199 (2000)
71. Filipowicz, W.: Evidence representation and reasoning in selected applications. In: Jdrzejowicz P., Nguyen, N.T., Hoang, K. (eds.) Lecture Notes in Artificial Intelligence, pp. 251–260. Springer-Verlag, Berlin (2011)
72. Moreno, R., Graa, M., Zulueta, E.: Rgb colour gradient following colour constancy preservation. Electron. Lett. **46**(13), 908–910 (2010)
73. Moreno, R., D'Anjou, A.: Hyperspectral image segmentation by t-watershed and hyperspherical coordinates. In: Graa, M., Toro, C., Posada, J., Howlett, R.J., Jain, L.C. (eds.) KES. Frontiers in Artificial Intelligence and Applications, vol. 243. pp. 2114–2121. IOS Press (2012)
74. Moreno, R., D'Anjou, A.: Hyperspectral image segmentation by t-watershed and hyperspherical coordinates. In: Graña, M., Toro, C., Posada, J., Howlett, R.J., Jain, L.C. (eds.) KES. Frontiers in Artificial Intelligence and Applications. vol. 243, pp. 2114–2121. IOS Press (2012)
75. Tweedale, J., Ichalkaranje, N., Sioutis, C., Jarvis, B., Consoli, A., Phillips-Wren, G.: Innovations in multi-agent systems. J. Netw. Comput. Appl. **30**(3), 1089–1115 (2006)
76. Tweedale, J.W., Jain, L.C.: Advances in information processing paradigms. In: Watanabe, T., Jain, L.C. (eds.) Innovations in Intelligent Machines-2. Studies in Computational Intelligence, vol. 376, pp. 1–20. Springer, Berlin (2012)
77. Barbucha, D., Czarnowski, I., Jędrzejowicz, P., Ratajczak-Ropel, E., Wierzbowska, I.: Influence of the working strategy on A-team performance. Smart Inform.Knowl. Manage. **206**, 83–102 (2010)(2010)
78. Wooldridge, M.: An Introduction to MultiAgent Systems, John Wiley & Sons (2002)
79. Friedman-Hill, E.: Jess in action: rule-based systems in Java. Manning Publications, Greenwich (2003)
80. Jedrzejowicz, P., Wierzbowska, I.: Impact of migration topologies on performance of teams of a-teams. In: Graña, M., Toro, C., Posada, J., Howlett, R.J., Jain, L.C. (eds.) KES. Frontiers in Artificial Intelligence and Applications, vol. 243, pp. 1161–1170. IOS Press (2012)
81. Barbucha, D.: An agent-based implementation of the multiple neighborhood search for the capacitated vehicle routing problem. In: Graña, M., Toro, C., Posada, J., Howlett, R.J., Jain, L.C., (eds.) Frontiers in Artificial Intelligence and Applications, KES vol. 243, pp. 1191–1200. IOS Press (2012)
82. Eglese, R.W.: Simulated annealing: a tool for operational research. Eur. J. Operat. Res. **46**, 271–281 (1990)
83. Voudouris, C., Tsang, E.: Guided local search and its application to the traveling salesman problem. Eur. J. Oper. Res. **113**, 469–499 (1999)
84. Gu, J., Huang, X.: Efficient local search with search space smoothing: a case study of the traveling salesman problem. IEEE Trans. Syst. Man Cybernet. **24**(5), 728–735 (1994)
85. Hansen, P., Mladenovic, N., Brimberg, J., and Moreno Perez, J.A.: Variable nighborhood search. In: Gendreau, M., and Potvin, J.-Y. (eds.) Handbook of Metaheuristics, International Series in Operations Research and Management Science, vol. 146, pp. 61–86. Springer, Berlin (2010)
86. Shaw, P.: Using constraint programming and local search methods to solve vehicle routing problems. In: Proceedings of Fourth International Conference on Principles and Practice of Constraint Programming CP-98. LNCS, vol. 1520, pp. 417–431 (1998)

87. Ropke, S., Pisinger, D.: An adaptive large neighborhood search heuristic for the pickup and delivery problem with time windows. Transport. Sci. **40**(4), 455–472 (2006)
88. Tedin, R., Becerra, J.A., Duro, R.J., Lede, I.M.: Towards automatic estimation of the body condition score of dairy cattle using hand-held images and active shape models. In: Graña, M., Toro, C., Posada, J., Howlett, R.J., Jain, L.C. (eds.) KES. Frontiers in Artificial Intelligence and Applications, vol. 243, pp. 2150–2159. IOS Press (2012)
89. Storn, R., Price, K.V.: Differential evolution - a simple and efficient heuristic for global optimization over continuous spaces. J. Global Optim. **11**, 341–359 (1997)
90. Caamaño, P., Tedín, R., Paz-Lopez, A., Becerra, J.A.: Jeaf: a java evolutionary algorithm framework. In: IEEE Congress on Evolutionary Computation, IEEE, pp. 1–8 (2010)
91. Guinard, D., Trifa, V., Wilde, E.: A resource oriented architecture for the web of things. In: Internet of Things (IOT), pp. 1–8 (2010)
92. Stirbu, V.: Towards a restful plug and play experience in the web of things. In: 2008 IEEE International Conference on Semantic Computing, pp. 512–517 (2008)
93. Borzemski, L., Kaminska-Chuchmala, A.: Knowledge engineering relating to spatial web performance forecasting with sequential gaussian simulation method. In: Graña, M., Toro, C., Posada, J., Howlett, R.J., Jain, L.C. (eds.): KES. Frontiers in Artificial Intelligence and Applications, vol. 243, pp. 1439–1448. IOS Press (2012)
94. Krige, D.: A statistical approach to some basic mine valuation problems on the Witwatersrand. J. Chem. Metall. Mining Soc. **52**, 119–139 (1951)
95. Borzemski, L.: The experimental design for data mining to discover web performance issues in a wide area network. Cybernet. Syst. **41**(1), 31–45 (2010)
96. Borzemski, L., Cichocki, L., Fraś, M., Kliber, M., Nowak, Z.: Mwing: A multiagent system for web site measurements. In: Nguyen, N.T. , Grzech, A., Howlett, R.J., Jain, L.C. (eds.) Agent and Multi-Agent Systems: Technologies and Applications, Lecture Notes in Computer Science, vol. 4496, pp. 278–287. Springer, Berlin (2007)
97. Ghosh, A., Tweedale, J.W., Nafalski, A., Dollard, M.: Multi-agent based system for analysing stress using the stresscafé. In: Graña, M., Toro, C., Posada, J., Howlett, R.J., Jain, L.C. (eds.) KES. Frontiers in Artificial Intelligence and Applications, vol. 243, pp. 1656–1665. IOS Press (2012)
98. Patterson, D.W.: Artificial neural networks theory and applications, Prentice Hall, International, pp. 247–264 (1996)
99. Rojas, R.: Neural Networks: A Systematic Introduction, ch. 2–6, ISBN 3-540-60505-3. Springer-Verlag, Berlin (1996)
100. Zang, Z., Zang, C.: Agent-based hybrid intelligent systems, LANI, vol. 2938, pp. 3–11. Springer-Verlag, Berlin (2004)
101. Jang, J.S.R., Sun, C.T., Mizutani, E.: Neuro-Fuzzy Soft Comput. A computational approach to learning and machine intelligence, Mathlab Curriculum Series (1997)

Chapter 2
Computing Efficiently Spectral-Spatial Classification of Hyperspectral Images on Commodity GPUs

Pablo Quesada-Barriuso, Francisco Argüello and Dora B. Heras

Abstract The high computational cost of the techniques for segmentation and classification of hyperspectral images makes them good candidates for parallel processing, in particular, for computing on Graphics Processing Units (GPUs). In this paper an efficient projection on the GPUs for the spectral–spatial classification of hyperspectral images using the Compute Unified Device Architecture (CUDA) for NVIDIA devices is presented. A watershed transform is applied after reducing the hyperspectral image to one band through the calculation of a morphological gradient, while the spectral classification is carried out by Support Vector Machine (SVMs). The results are combined with an adaptive majority vote. The different computational stages are concatenated in a pipeline that minimizes the data transfer between the main memory of the host computer and the global memory of the graphics device to maximize the computational throughput. The memory hierarchy and the thousands of threads available in this architecture are efficiently exploited. It is possible to study different data partitioning strategies and thread block arrangements in order to promote concurrent execution of a large number of threads. The objective is to efficiently exploit commodity hardware with the aim of achieving real-time execution for on-board processing.

Keywords Hyperspectral images · Watershed · Classification · CUDA

P. Quesada-Barriuso (✉) · F. Argüello · D. B. Heras
Centro de Investigación en Tecnoloxías da Información (CITIUS), University of Santiago de Compostela, Rúa de Jenaro de la Fuente Domínguez, Santiago de Compostela 15842, Spain
e-mail: pablo.quesada@usc.es

F. Argüello
e-mail: francisco.arguello@usc.es

D. B. Heras
e-mail: dora.blanco@usc.es

J. W. Tweedale and L. C. Jain (eds.), *Recent Advances in Knowledge-based Paradigms and Applications*, Advances in Intelligent Systems and Computing 234,
DOI: 10.1007/978-3-319-01649-8_2,

2.1 Introduction

Recent advances in sensor technology have led to hyperspectral images being now widely available [1, 2]. The special characteristics of hyperspectral images, which provide a detailed spectrum for each pixel, allow distinguishing among physical materials and objects even at pixel level, presenting new challenges to spectral analysis, target detection, image segmentation or classification. Nevertheless, the large number of spectral channels of the hyperspectral images makes most of the commonly used methods designed for the processing of grey level or color images not appropriate. To take full advantage of the rich information provided by the spectral dimension new algorithms are required.

The supervised classification of hyperspectral images has been a very common topic in the last decades. Pixel-wise classifiers, for instance, consider only the spectral information of the pixel [1, 3–5]. In particular, pixel-wise classification by Support Vector Machine (SVM) classifiers has been introduced and shown good results when a small number of training samples are available [3]. However, this pixel-wise classification does not consider information about spatial structures. Therefore, the classification can also take advantage of the spatial relationships among pixels, allowing more elaborate spectral–spatial models for a more accurate segmentation and classification of the image [6–8]. The spatial information can be included considering different approaches. The first approach consists in including information from the closest neighborhood of a pixel through morphological filtering [9], morphological leveling [6] or Markov random fields [10]. The second approach consists in carrying out a segmentation of the image by methods that are usually based in graphs [11]. Among these some unsupervised methods have been widely used: partitional clustering [7], hierarchical segmentation [12], MSF [13] and watershed [8]. The watershed transform is a widely used method for non-supervised image segmentation, specially suitable for low contrast images [14]. It is usually applied to the morphological gradient of a two dimensional image for extracting homogeneous regions with respect to grey level values.

Recently, Tarabalka et al. [8] have presented a spectral–spatial classification scheme for hyperspectral images that uses the watershed transform. It is based on an SVM spectral classification, followed by a Majority Vote (MV) process among the classified pixels within the same watershed region. Among the proposals presented by the authors to reduce the image to one band, such as multidimensional or vectorial gradients. One of the most efficient approaches, in terms of classification quality, is obtained through a Robust Color Morphological Gradient (RCMG) calculation. The good classification results of this proposal in urban and open areas had led us to adopt it in this work.

The computational cost of the techniques for segmentation and classification of hyperspectral images is high, which makes them good candidates for parallel and, in particular, for General-Purpose Computing on Graphics Processing Units (GPGPUs). The focus of this study is to provide a solution for a GPU platform, adapting the hyperspectral processing to a low cost parallel computing architecture. With this approach the on-board processing of information is possible without the need for bulky high performance computing infrastructures.

In most cases neither sequential nor existing parallel algorithms can be directly implemented in the GPU and it is necessary to modify the flow of the computations in order to fully exploit the architecture. The use of Graphics Processing Units to process hyperspectral images has been gaining popularity in recent years. For instance, algorithms for spectral unmixing [15, 16], target detection [17, 18], classification [1] and segmentation [19, 20] have led to more complete tools [21, 22]. A spectral–spatial GPU classification tool was presented in [22]. In this tool the spatial information is introduced by MV within a fixed window where each pixel is assigned to the most predominant class, so the spatial structure of the image is not fully considered. As a result, this MV implementation may generate different classes within the same watershed region, unlike in [8].

The interest is on exploring GPU architectures for hyperspectral processing by developing techniques that can be efficiently projected on GPU consumer platforms with the objective of achieving real-time execution that makes on-board processing possible. In this paper a spectral–spatial classification scheme for hyperspectral images based on [8] is presented, specially adapted for GPU processing using CUDA. The process consists in the calculation of a morphological gradient operator, that reduces the dimensionality of the hyperspectral image, followed by the calculation of a watershed transform based on Cellular Automata (CA) over the resulting 2D image, and a spectral classification based on SVM. A MV process combines the spectral and spatial results. The thousands of threads available in the GPU are efficiently exploited. The different stages are concatenated in a pipeline processing that minimizes the data transfers between the host and the device and maximizes the computational throughput. Furthermore, data are reused within the GPU, taking advantage of the shared memory and cache hierarchy of the architecture. In addition, different hyperspectral data partitioning strategies and thread block arrangements are studied in order to effectively exploit the memory and computing capabilities of the GPU architecture.

The reminder of this paper is organized as follows: in Sect. 2.2 some GPU and CUDA fundamentals are introduced. Section 2.3.1 introduces the morphological gradient, Sect. 2.3.2, the watershed transform, and Sect. 2.3.3 the majority vote approach for spectral–spatial classification. The implementations of the algorithms and the results obtained are discussed in Sects. 2.4 and 2.5, respectively. Finally, Sect. 2.6 presents the final remarks.

2.2 GPU Architecture

The most recent GPUs provide massively parallel processing capabilities based on a data parallel architecture. The NVIDIA GPU architecture is organized into a set of Streaming Multiprocessors (SMs), each one with many cores called streaming processors [23], as shown in Fig. 2.1a. These cores can manage hundreds of threads

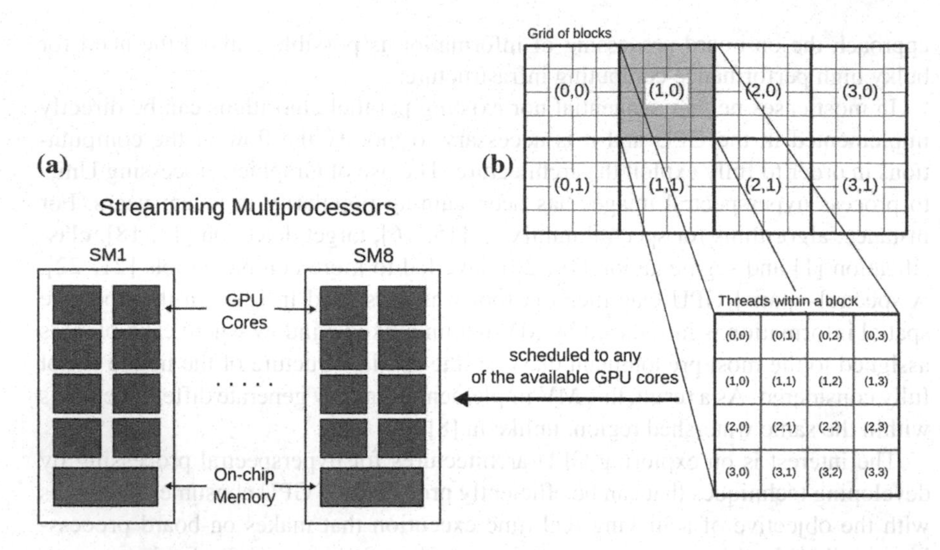

Fig. 2.1 NVIDIA CUDA architecture. (**a**) Streaming multiprocessors and (**b**) organizations of Grid, blocks and threads

in a Single Instruction Multiple Data (SIMD) programming model. The GPU cores execute the same instruction simultaneously on different data unlike the multicore processors that are Multiple Instruction Multiple Data (MIMD) (different cores execute different threads operating on different data).

CUDA for NVIDIA devices, is an Application Programming Interface (API) for writing programs that are executed in the GPU. A CUDA program, which is called a kernel, is executed by thousands of threads grouped into blocks, as illustrated in Fig. 2.1b. The Compute Unified Device Architecture (CUDA) has a global memory of Dynamic Random Access Memory (DRAM) that is available for all the blocks. There is also an on-chip *shared memory* space only available per block. This feature enables an extremely rapid read/write access to the data in this memory but with the lifetime of the block. Furthermore, it is not possible to read or write data to the shared memory allocated to another block. Finally, each thread has its own *local memory* and *registers*. Examples include the NVIDIA G80 and GT200 graphics cards series.

The Fermi and Kepler architectures [24] have also a *cache hierarchy* consisting of a configurable L1 and a unified L2 caches. The 64 KB of on-chip memory can be configured as 48 KB of shared memory and 16 KB of L1 cache or vice versa. There are 64 KB of this memory available for each SM. The L2 is a unified cache up to 1,536 KB shared by all the SMs. The accesses to the DRAM are cached in this memory hierarchy. The NVIDIA Tesla GF100 and the GeForce 500 series are examples of the Fermi architecture. The Tesla K-series family of products includes the Kepler K10, K20 and K20X GPU accelerators with different chipsets. In particular the Tesla K20X based on the GK110 chipset incorporates 2688 CUDA cores and 6 GB of memory. These chipsets can be found in commodity GPUs like the GTX680 graphics card, used in this work, which has a GK104 chipset (1536 CUDA cores and 2 GB of memory).

The challenge of GPU programming is to increase the computational throughput. To achieve this, important aspects that must be considered are [25]: minimizing CPU–GPU data transfers, aligning accesses to consecutive memory locations, maximizing data reuse, balancing the workload among threads, and minimizing their divergence.

2.3 Spectral–Spatial Classification of Hyperspectral Images

Hyperspectral images are basically digital pictures where each pixel is represented by a set of n values. Each value corresponds to a spectral component across the visible and infrared light bands [18]. The number of captured bands depends on the properties of the hyperspectral sensor. For example, the well known Reflective Optics System Imaging Spectrometer (ROSIS) is able to record 103 spectral bands [26], while the Airbone Visible-Infrared Imaging Spectrometer (AVIRIS) is able to record 224 spectral bands [27].

Most classification methods for hyperspectral images process each pixel independently using pixel-wise classifiers, but do not take into account the spatial information of the neighborhood [28]. Nevertheless, it has been proved that the classification results significantly improve when spatial information is incorporated [6–8].

An efficient approach to integrate spectral and spatial information in a classification system is defined by Tarabalka et al. [8]. The process consists of the stages shown in Fig. 2.2. On one hand, the spectral processing is applied over the hyperspectral

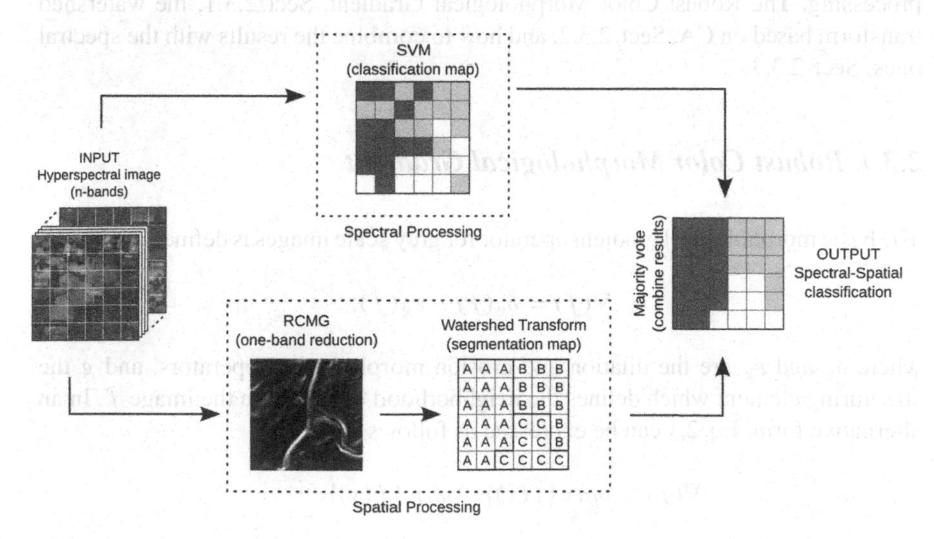

Fig. 2.2 Spectral-spatial classification scheme, which consists of a spectral stage (*top*), a spatial stage (*bottom*), and a final stage to combine the results

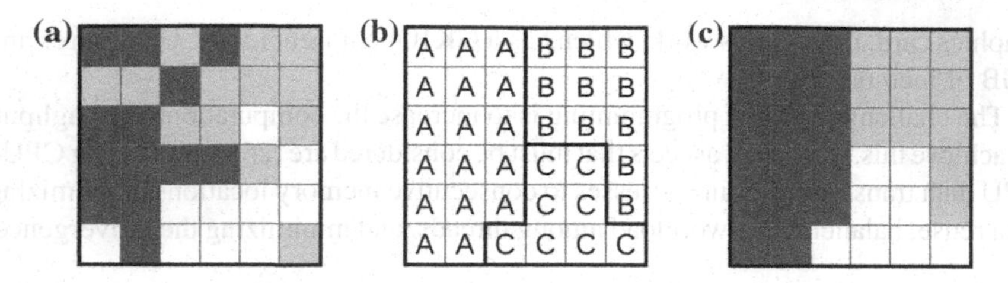

Fig. 2.3 Example of majority vote application for spectral–spatial classification. (**a**) Classification map; (**b**) Segmentation map; (**c**) Majority vote within a segmented region

input image using a SVM that produces a classification map (shown at the top of the figure). Each pixel of this map belongs to one class predicted by the SVM (three classes in this example). On the other hand, the spatial processing, applied to the one–band image generated after a RCMG calculation, creates a segmentation map using a watershed transform (shown in the bottom of the figure). In this map, all of the pixels are labelled according to the region they belongs to.

Finally, the spectral and spatial results are combined using a majority vote process. Each pixel in a watershed region is assigned to the most predominant class among the classes within the same region. The output of this scheme, as shown in Fig. 2.2, is a more accurate hyperspectral classification of the image compared to a standalone spectral classification. The procedure for combining the results is illustrated in detail in Fig. 2.3 for the case of three spectral classes, represented as three colors in Fig. 2.3a. The segmentation map, with regions A, B, C, and the results of the MV are displayed in Fig. 2.3b, c.

In the following sections we explain in detail the different steps of the spatial processing. The Robust Color Morphological Gradient, Sect. 2.3.1, the watershed transform based on CA, Sect. 2.3.2, and how to combine the results with the spectral ones, Sect. 2.3.3.

2.3.1 Robust Color Morphological Gradient

The basic morphological gradient operator for grey scale images is defined as Eq. 2.1:

$$\nabla(f) = \delta_g(f) - \varepsilon_g(f), \tag{2.1}$$

where δ_g and ε_g are the dilation and erosion morphological operators, and g the structuring element which defines the neighborhood of a pixel in the image f. In an alternative form, Eq. 2.1 can be expressed as follows in Eq. 2.2:

$$\begin{aligned}\nabla(f) &= \underset{x\in g}{max}\,\{f(x)\} - \underset{y\in g}{min}\,\{f(y)\} \\ &= \max(|f(x) - f(y)|) \quad \forall x, y \in g,\end{aligned} \tag{2.2}$$

giving the greatest intensity difference between any two pixels within the structuring element. In this way Eq. 2.2 can easily be extended to color images [29], which have a pixel vector of three components, i.e. the red, green and blue channels of color.

Let $\mathbf{x}$ be a pixel vector of a color image and $\chi = [\mathbf{x}_1, \mathbf{x}_2, \ldots, \mathbf{x}_n]$ be a set of n pixel vectors in the neighborhood of $\mathbf{x}$, and the set χ contains $\mathbf{x}$. The Color Morphological Gradient (CMG), $\nabla(\mathbf{f})$, using the Euclidean distance, is defined as Eq. 2.3:

$$\nabla(\mathbf{f}) = \max_{i,j \in \chi} \{||\mathbf{x}_i - \mathbf{x}_j||_2\}, \tag{2.3}$$

whose response is the maximum of the distances between all pairs of vectors in the set χ. As the CMG is very sensitive to noise and may produce edges that are not representative of the gradient, a RCMG is proposed in [29], based on pairwise pixel rejection of Eq. 2.3. The RCMG , $\nabla(\mathbf{f})_{Robust}$, is defined as Eq. 2.4:

$$\nabla(\mathbf{f})_{\text{Robust}} = \max_{i,j \in \chi - R_s} \{||\mathbf{x}_i - \mathbf{x}_j||_2\}, \tag{2.4}$$

where R_s is the set of s pairs of pixel vectors removed. The pairs removed are those that are furthest apart. The RCMG is therefore a vectorial gradient operator based on the Euclidean distances of pixel vectors.

A pixel vector also refers to a pixel of the hyperspectral image with all the n-bands as components of a n-dimensional vector. Thus, using the RCMG, a hyperspectral image may be reduced to a single band and be used as input for the watershed transform.

Regarding GPU concerns, the calculation of Eq. 2.4 is split into partial operations and then the partial results are combined to find the maximum distance. There are two possible work distribution strategies among thread blocks which will be explained in Sects. 2.4.1.1 and 2.4.1.2.

2.3.2 Watershed Transform Based on Cellular Automata

Regarding segmentation, a watershed transform based on CA is applied, because of the simplicity of the computing model of the CA that can model complex problems easily, and because the computations for pixels are highly independent, and thus very adequate for streaming parallel processing architectures like the GPU.

The watershed algorithm is a widely used method for non-supervised image segmentation, specially suitable for low-contrast images [14]. If a grey scale image is represented as a topographic relief, where the height of each pixel is directly related to its grey level, the dividing lines of the basins of attraction of rain falling over the regions are called watershed lines [14]. Various definitions, algorithms and implementations can be found in the literature [30]. In this paper the Hill-Climbing algorithm based on the topographical distance by Meyer is adopted [31]. This algorithm starts by detecting and labelling all minima in the image with unique labels.

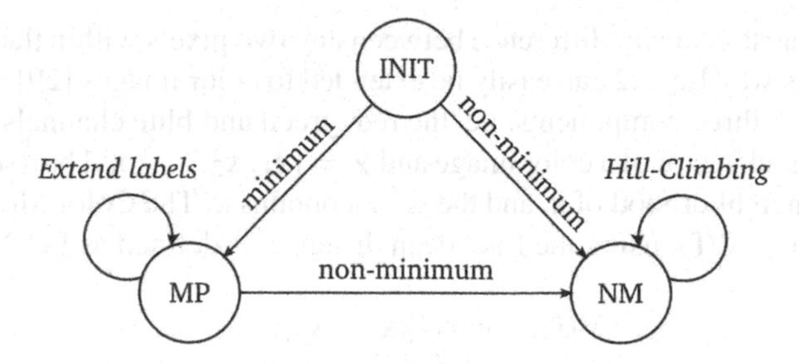

Fig. 2.4 Three-state automation implementing the Hill-Climbing algorithm [32]

Then, the labels are propagated upwards, *climbing up the hill*, following the path defined by the lower slope of each pixel. At the end, all pixels have a label that identifies the region to which they belong. No implicit lines are generated with this algorithm, so the watershed lines are the limits between these regions.

CA are computing models composed of a set of cells arranged in a regular grid, with each cell connected to its adjacent neighbors. The CA evolve in discrete time steps, according to a collection of states and a set of transition rules. One of the main characteristics of CA is that updates are made for each cell considering only local information, so the concept of parallelism and, in particular, streaming processing, is implicit in the automata. The updates of the cells can be carried out synchronously or asynchronously [33]. In the latter case, the grid can be partitioned into different regions which can be independently updated an unbounded number of times.

Galilée et al. proposed a three-state cellular automaton implementing the Hill-Climbing algorithm [32] that is shown in Fig. 2.4 (MP stands for Minimum or Plateau state and NM for Non-minimum state). The main advantage of this Watershed Transform based on Cellular Automata (CA–Watershed) is that minima detection, labeling, and climbing the steepest paths are simultaneously and locally performed.

Each cell of the automaton computes a pixel of the image. First, the pixels are sequentially labelled and the state of each pixel is initialized to one of two possible states. Considering that a plateau is a region of constant grey value within the image, these states are MP and NM. If a pixel is within a plateau, it switches to the MP state. Otherwise, the state of the pixel switches to NM. Figure 2.5a shows an example of a 1D image represented as a terrain (lines) and the corresponding grey values of each pixel (squares). The numbers within each square in this figure are initial label values.

Once the pixels have been initialized, the following steps update the automaton. This is an iterative task that processes the MP and NM states as follows: The pixels of a plateau, i.e. MP state, extend the label with the minimum value along the pixels belonging to that plateau, in case of a plateau that is minimum as indicated in Fig. 2.5b, and change their state if the plateau is non-minimum. If the state of the pixel is NM, the label is propagated through the lower slope as shown in Fig. 2.5c, where labels 3 and 9 are being propagated upwards, climbing up the hills. This iterative task ends

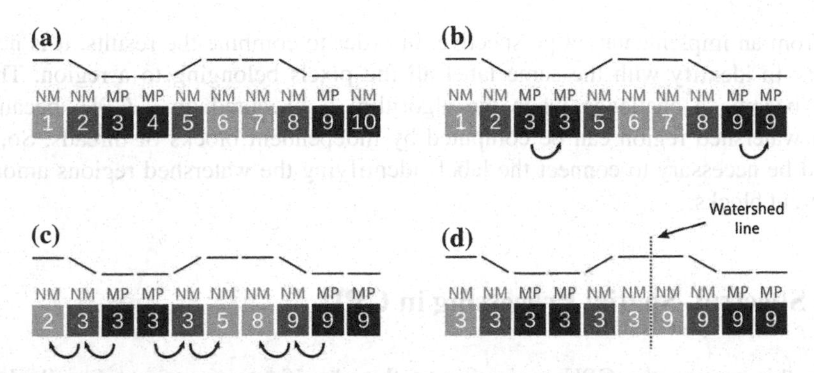

Fig. 2.5 1D image represented as a terrain (*top lines*) and the corresponding *grey* values of each pixel (*bottom squares*). (**a**) Init state, (**b**) MP update, (**c**) NM update, (**d**) final segmentation

when no more changes occur as in Fig. 2.5d. The result is a segmentation map where each region is represented by the label corresponding to the seed pixel that generated the region. The watershed lines can be later defined as the borders among regions.

The CA–Watershed can be synchronously or asynchronously implemented. The asynchronous implementation is non-deterministic and may lead to different segmentation results. A formal proof of correctness and convergence towards a watershed segmentation using a mathematical model of data propagation in a graph is presented in [32].

The asynchronous algorithm is particularly suitable for the CUDA computing model as it was shown by Quesada-Barriuso et al. [34]. Different regions of the image can be simultaneously and independently updated during certain number of steps, thus reducing the number of points of synchronization, so the exploitation of parallelism is maximized.

2.3.3 Majority Vote

The MV is a process for determining which out of an arbitrary number of candidates has received most votes, considering a vote as a particular property or attribute. One possible implementation of MV takes as input an array with the votes for each candidate, and returns the element with most votes after one pass over the whole vector [35]. In the hyperspectral classification context, the MV within a fixed window, i.e. fixed neighborhood, is a standard spatial regularization procedure when it is applied after a pixel-wise classification [8]. However, using the regions created by the segmentation process as in this work, i.e. using an adaptive neighborhood, the spatial structures that may be present in the image are taken into account in a more realistic way. So, using an adaptive neighborhood, the MV process integrates the spectral and the spatial information that are available per pixel within each watershed region, summing up the votes that identify the spectral class for each pixel [36].

From an implementation perspective, in order to combine the results, it is necessary to identify with the same label all the pixels belonging to a region. This may become a challenge when the algorithm is executed on a GPU, because each watershed region can be computed by independent blocks of threads. So, it could be necessary to connect the labels identifying the watershed regions among different blocks.

2.4 Spectral–Spatial Processing in GPU

In this section the GPU projection of the classification process described in Sect. 2.3 is detailed. The hyperspectral image must be divided into regions that are distributed among the thread blocks. The regions will be one, two or three dimensional depending on the executed stage, enabling all the threads to perform useful work, and therefore exploiting the thousands of threads available in the GPU.

For the RCMG and the CA–Watershed stages, each data region must be extended with a border of size one because the processing of each pixel requires data of its neighbors. As an example, Fig. 2.6a shows an image divided into 4×4 pixel regions assigned to blocks of 4×4 threads, and Fig. 2.6b the extended region for one of the blocks. Threads on the edge of the block must perform extra work loading the data corresponding to the border. In practice, rectangular regions are considered. Using a rectangular block, with the longest dimension being the one along which data is stored in global memory, the data of the border is packed in the minimum number of cache lines. This way the overhead associated to global memory accesses is minimized [37].

Thanks to the pipeline processing, the number of computations and the required bandwidth are reduced in the majority vote stage, because all the pixels in the same watershed region are already identified by the same watershed label. So, there is no need to create new data structures and copy them to the GPU memory.

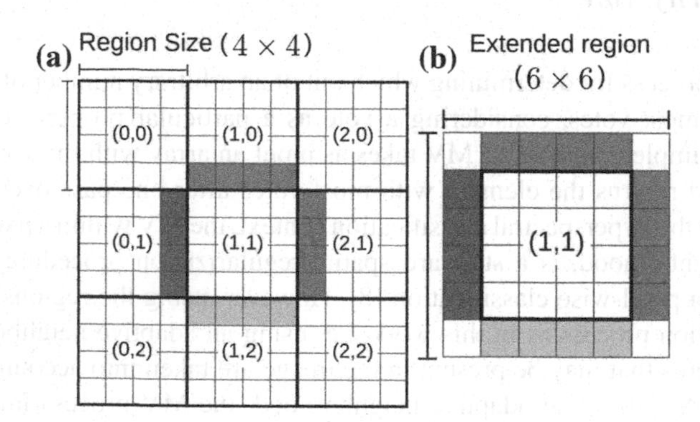

Fig. 2.6 An image divided into, 4×4 pixel regions and (**b**) the extended region for one of the blocks

Regarding the spectral processing, different implementations in GPU of SVM are available in the literature [38–42]. Among those that provide the source code performing training and classification, and producing a final classification map, the selected library is the GPUSVM by Catanzaro et al. [40]. This implementation considers the standard two-class soft-margin SVM classification problem. With the use of the CUDA Basic Linear Algebra Subroutines (CUBLAS)[1] to perform the classification, the library takes maximum profit from the latest CUDA releases.

2.4.1 Robust Color Morphological Gradient

The workflow of the RCMG algorithm, summarized in Fig. 2.7 is divided into three steps. First, for all the pixel vectors, threads within a block cooperate to obtain the distances of the set χ, for calculating Eq. 2.3, and computing the CMG. Second, the pair of pixels R_s that are furthest apart, required for Eq. 2.4, are found and the RCMG is calculated with the remaining distances in the third step. So, finally a one-band gradient is obtained. In this work $R_s = 1$, i.e. only one pair of pixels is removed.

The hyperspectral image can be partitioned in the spatial or the spectral domains. From a processing point of view, two different algorithms have been implemented. One based on *spatial partitioning* within a block, as shown in Fig. 2.8 a and described in Sect. 2.4.1.1. Another based on *spectral partitioning* within a block, described in Sect. 2.4.1.2 and shown in Fig. 2.8b. In both cases, the input image is stored in global memory so that consecutive threads access consecutive global memory locations. The intermediate results that are necessary in order to calculate the distances are stored in shared memory.

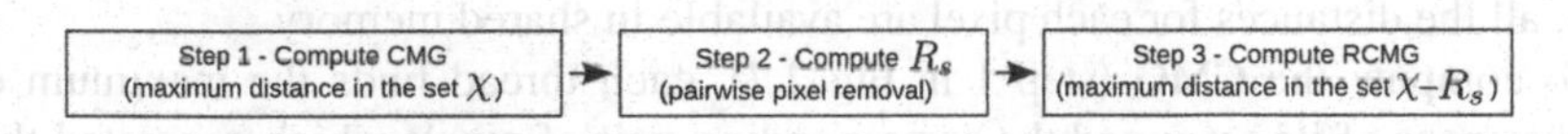

Fig. 2.7 RCMG algorithm work-flow

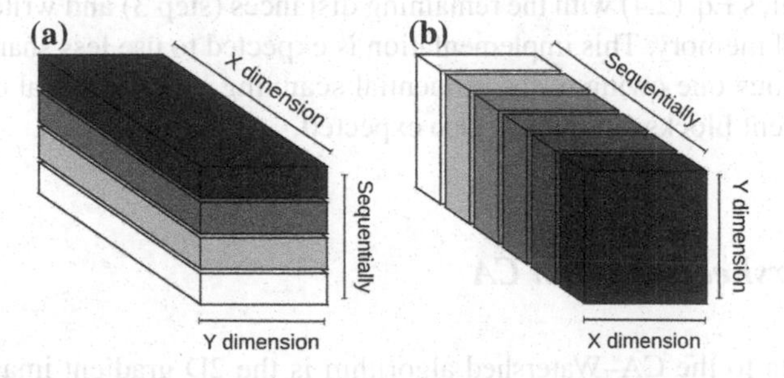

Fig. 2.8 Kernel configuration for spatial (**a**) and spectral (**b**) partitioning

[1] See CUBLAS at https://developer.nvidia.com/cublas

2.4.1.1 Spatial Partitioning RCMG

In this implementation, each thread processes one spectral component, as shown in Fig. 2.8a, and a group of threads cooperate in a reduction operation, where the largest dimension of a thread block indexes the different spectral components. For each region of the image, all the spectral components of each pixel vector are consecutively stored in global memory. The kernel is configured to work in blocks of $x \times y$ threads, corresponding to the X and Y dimensions of Fig. 2.8a. For each block, all the threads load different components of each pixel vector simultaneously and compute a partial result $(\mathbf{x}_i)^2 - (\mathbf{x}_j)^2$ of Eq. 2.3. Then, the threads in the X dimension cooperate in a reduction operation [43] for computing the CMG (step 1). Half of the threads work in the reduction, and the number of active threads is halved at each iteration as the reduction proceeds.

One thread in the X dimension finds the pair of pixels that generated the maximum distance (step 2) and computes the RCMG (step 3) with the remaining distances. Finally, the RCMG is written in global memory.

2.4.1.2 Spectral Partitioning RCMG

In the spectral partitioning RCMG, each thread processes all the spectral components of a pixel, as shown in Fig. 2.8b. For each region, data are stored in row-major order for each band. The kernel is configured to work in blocks of 32×4 threads corresponding to the X and Y dimensions in Fig. 2.8b. Threads within a block process a region of each spectral band in a loop through all the bands (sequential processing). At each iteration, data corresponding to a new band are loaded in shared memory, and the partial results $(\mathbf{x}_i)^2 - (\mathbf{x}_j)^2$ of Eq. 2.3 are computed and stored. At the end of the loop, all the distances for each pixel are available in shared memory.

To compute the CMG (step 1 in Fig. 2.7), each thread finds the maximum of the distances of its set χ and the corresponding pair of pixels which generated that maximum. Having identified the two pixel vectors that are furthest apart (step 2), each thread computes Eq. (2.4) with the remaining distances (step 3) and writes the result back to global memory. This implementation is expected to use less shared memory that the previous one owing to the sequential scanning in the spectral domain. So, more concurrent blocks per SM are also expected.

2.4.2 *Watershed Based on CA*

The input data to the CA–Watershed algorithm is the 2D gradient image obtained from the RCMG algorithm. The CA–Watershed can be asynchronously implemented as it was mentioned in Sect. 2.3.2, which is up to four times faster than the CUDA synchronous implementation [34]. In this section the asynchronous algorithm is described, which has the advantage of reusing information within a block, efficiently exploiting the shared and cache memories of the GPU.

The algorithm has two kernels implementing the initialization and updating stages of the CA–Watershed. These kernels are configured to work in blocks of 32×4 threads operating over 32×4 pixel regions of the image. Data structures have been compressed in order to reduce the storage requirements to 8 bytes per pixel as in [34]. With the first kernel, the automaton is initialized. Once all the data have been initialized, they are packed into 8 bytes per pixel before transferring them to global memory.

The updating stage is a hybrid iterative process that includes intra-block updates and inter-block updates. Each region is synchronously updated, for instance all cells within a region are updated at each time step, while the regions themselves are asynchronously updated (an update of all the blocks is performed at each inter-block step).

During the intra-block updating the values used from outside the block (a border of size one) are kept constant and equal to their values at the beginning of the stage. In the inter-block updating process, data are read at the block borders, which allows the data propagation across the entire grid.

On each call to the CUDA kernel, an inter-block update takes place, where each step is a set of intra-block updates. For each block, once data are loaded in shared memory from an input buffer, the pixels are modified in registers according to their state, and stored back to shared memory in an iterative intra-block process within each region.

The intra-block updating ends when no new modifications are made with the available data within the region. Then the data in shared memory are packed and stored in global memory in an output buffer. This operation is repeated several times in an iterative inter-block process. The algorithm ends when all regions have been flooded and each pixel is labelled with a value indicating the region it belongs to.

The CA–Watershed implementation not only exploits efficiently the resources of the GPU, as the shared memory, but also generates a segmentation map where the pixels are connected. Figure 2.9a shows an example of an image segmented into three regions, represented as “A”, “B”, “C”. The grey lines in each region indicate that after the segmentation process every pixel of each region has the same label,

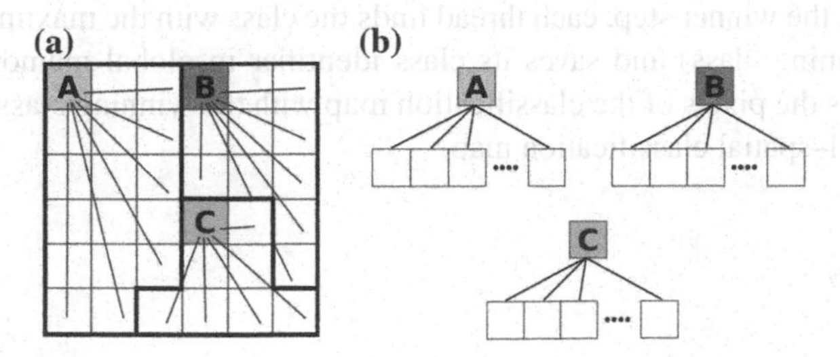

Fig. 2.9 An example of an image segmented into three regions, **(a)** represented as “A”, “B”, “C”, and **(b)** the connected components created from the labels

that of the pixel from which the region was created. So, the pixels are connected as shown in Fig. 2.9b without the need of performing any component labelling process [44]. Thus, the output of this algorithm can be used directly in the final stage of the spectral–spatial classification scheme.

2.4.3 Majority Vote

The MV, when applied to this hyperspectral classification, processes the pixels within each segmented region. In this implementation, a region can be assigned to different thread blocks, therefore all the pixels belonging to the same region must be connected, as shown in Fig. 2.9b.

By using the segmentation map, such as the one in Fig. 2.9a, the pixels of each watershed region are already connected, so the MV can be projected in the GPU following the steps: *voting, winner and updating*. The voting step counts the number of SVM classes for each region. The winner step finds the class with the maximum number of votes per region, and finally, the updating step assigns the winner class to the pixels within the region. Each step is performed by a separate kernel that is configured to work in one dimensional blocks of threads. In the first and third kernel each thread operates on one pixel of the image, while in the second kernel each pixel operates on the information collected for one region of the segmentation map. So, for the second kernel less blocks need to be executed.

One majority vote per watershed region is performed. As the number of these regions is unknown a priori, the first approach would be to allocate in global memory data structures of a large enough size to compute as many regions as pixels in the image. With the aim of saving memory resources, the number of regions generated by the CA–Watershed algorithm are calculated prior to the voting step. Once the number of regions is known, a two dimensional data structure is defined in global memory being the number of watershed regions and the number of spectral classes the dimensions of the structure.

For the voting kernel, each thread adds one vote to the corresponding class, using the label as an index to reference its region. As two or more threads can vote in the same region to the same class with no predictable order, the voting is done by atomic operations. In the winner step, each thread finds the class with the maximum number of votes (winning class) and saves its class identifier in global memory. The last kernel updates the pixels of the classification map with the winning class, producing a new spectral–spatial classification map.

2.5 Results

The algorithms have been evaluated on a PC based in the Nehalem microarchitecture with an Intel quad-core i7-860 microprocessor (8 MB Cache, 2.80 GHz) and 8 GB of Double Data Rate type three (DDR3) Synchronous DRAM. The code has been

Table 2.1 Classification accuracy as percentages for the SVM and the whole spectral–spatial scheme in CPU and GPU for the hyperspectral image of Pavia in terms of OA, AA and CS

		SVM CPU	SVM GPU	Spect–Spat. CPU	Spect–Spat. GPU
OA		89.77	89.78	94.55	**94.63**
AA		91.49	91.50	**95.00**	94.97
CS	RO				
Asphalt	0.083	84.80	84.83	**94.74**	94.59
Meadows	0.029	90.37	90.39	94.89	**95.12**
Gravel	0.187	78.75	78.85	**86.37**	85.66
Trees	0.171	**96.57**	**96.57**	93.93	93.90
Metal sheets	0.197	99.55	99.55	**99.63**	**99.63**
Bare Soil	0.106	88.51	88.53	92.86	**93.30**
Bitumen	0.282	95.04	95.04	96.69	**96.84**
Bricks	0.140	89.90	89.90	**95.98**	95.79
Shadows	0.244	**99.89**	**99.89**	**99.89**	**99.89**

compiled using gcc version 4.4.3 with OpenMP 3.0 support under Linux. For the CUDA implementation we run the algorithms on a NVIDIA Kepler architecture with the GK104 chipset (1536 CUDA cores and 2 GB of Graphics Double Data Rate type five (GDDR5) Synchronous DRAM). The GPU is a GeForce GTX680 with 64 KB of on-chip memory that can be distributed among L1 cache and shared memory and 8 SMs which can execute up to 16 concurrent blocks giving a total maximum of 2048 threads per SM. The CUDA code has been compiled using nvcc and the 4.2 toolkit, also under Linux.

The results are expressed in terms of execution times and speedups. For the SVM spectral classification the speedups are calculated with respect to the LIBSVM [45] that is a sequential library. For the remaining steps of the spectral–spatial classification scheme of Fig. 2.2, the reference codes in CPU are optimized OpenMP parallel implementations considering 4 threads because four cores are available in the Intel Core i7. The tests were run on two hyperspectral airborne images that were obtained from the Basque University (UPV/EHU)[2]: A 103-band ROSIS image from the *University of Pavia* (Pavia) with a spatial dimension of 610×340 pixels, and a 204-band AVIRIS image of 512×217 pixels taken over the *Salinas Valley, California* (Salinas). Although both images are of approximately the same global size, Pavia is larger in the spatial domain while Salinas is larger in the spectral domain.

The final results were compared to the available ground truth of each image. these results are validated using the Overall Accuracy (OA), which is the percentage of correctly classified pixels in the whole image, the Class Accuracy (CS), which is the percentage of correctly classified pixels for a given class, and the Average Accuracy (AA), which is the mean of the CS for all the classes [13]. Tables 2.1 and 2.2 shows

[2] Hyperspectral Remote Sensing Scenes available at http://www.ehu.es/ccwintco/index.php/Hyperspectral_Remote_Sensing_Scenes

Table 2.2 Classification accuracy as percentages for the SVM and spectral–spatial scheme in CPU and GPU for the hyperspectral image of Salinas in terms of OA, AA and CS

		SVM CPU	SVM GPU	Spect–Spat. CPU	Spect–Spat. GPU
OA		93.55	93.45	**94.43**	94.37
AA		96.82	96.76	**96.90**	96.89
CS	RO				
Brocoli_green_weeds_1	0.100	99.75	99.75	99.75	**99.80**
Brocoli_green_weeds_2	0.100	99.79	99.76	**100.00**	**100.00**
Fallow	0.100	99.85	99.85	**100.00**	**100.00**
Fallow_rough_plow	0.100	99.71	99.71	**99.78**	**99.78**
Fallow_smooth	0.100	98.77	98.77	**99.14**	**99.14**
Stubble	0.100	99.65	99.65	**99.85**	**99.85**
Celery	0.100	99.62	**99.92**	99.80	99.80
Grapes_untrained	0.100	89.64	89.57	**93.42**	93.12
Soil_vinyard_develop	0.100	99.95	99.81	**99.98**	**99.98**
Corn_senesced_green_weeds	0.100	98.05	97.53	**98.78**	98.72
Lettuce_romaine_4wk	0.100	**98.97**	**98.97**	**98.97**	**98.97**
Lettuce_romaine_5wk	0.100	**99.79**	**99.79**	99.58	**99.79**
Lettuce_romaine_6wk	0.100	**99.67**	**99.67**	95.96	95.74
Lettuce_romaine_7wk	0.100	**95.51**	95.42	94.77	94.77
Vinyard_untrained	0.100	71.05	70.82	**71.68**	71.66
Vinyard_vertical_trellis	0.100	**99.11**	**99.11**	98.95	99.17

the OA, AA, and CS percentages for the SVM and the spectral–spatial classification scheme in CPU and GPU for the images of Pavia and Salinas. RO stands for the ratio between the number of training samples and the number of testing samples for each class. The best accuracies are indicated in bold. These results are similar to those published in [6–8] when combining spectral and spatial information. Overall, the image of Pavia gives the best results in terms of spectral–spatial classification with an OA improvement of 4.85 points over the SVM. Similar results are obtained in CPU and GPU. The image of Salinas has a very high OA score with the SVM classification and thus, less room for improvement with the spectral–spatial scheme. The improvement for this image is 0.92.

Figures 2.10 and 2.11 show from left to right the SVM classification map, the RCMG results, the segmentation map represented as watershed lines, and the majority vote for the University of Pavia and the Salinas Valley, respectively. The number of watershed regions are 22,678 for the first image and 8,423 for latter one. An over-segmented result is observed in Pavia, while the number of regions in Salinas is smaller due to the larger plateaus present in that image

The performance results are summarized in Table 2.3 for Pavia and Table 2.4 for Salinas. The time to transfer the hyperspectral image from CPU to the GPU global memory at the beginning is included in the spectral stage. The data transfer time for copying the final results back to the CPU is 0.6 ms for Pavia and 0.3 ms for

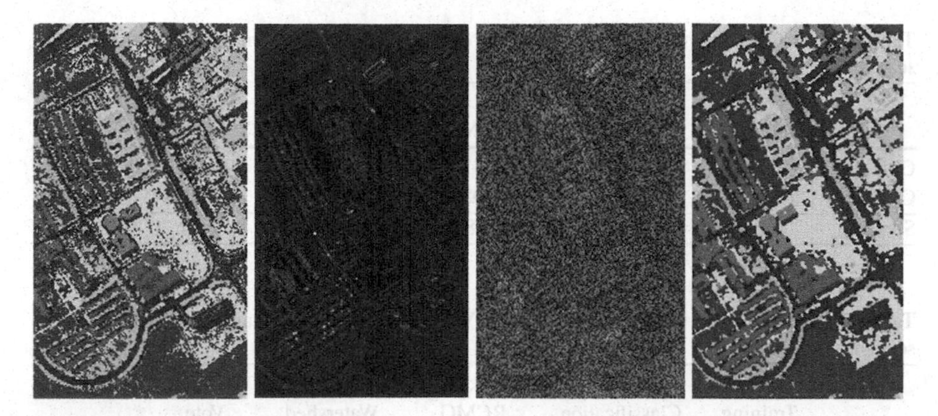

Fig. 2.10 From *left* to *right*, the GPUSVM classification map, the RCMG, CA–Watershed *lines* imposed over a *false color* composition to assist in visualizing the segmentation map, and the final classification by majority vote, of the hyperspectral image of Pavia

Fig. 2.11 From *left* to *right*, the GPUSVM classification map, the RCMG, the CA–Watershed *lines* and the final classification by majority vote, of the hyperspectral image of Salinas

Salinas, resulting in less than 0.003 % of the total time. The total times indicate that, even with the high speedups obtained, the times required in GPU are around 17 s for Pavia and 59 s for Salinas. These values are far from real-time, mainly due to the cost of the spectral classification, that accounts for the 81.2 % of this GPU time. So the real-time objective can only be achieved if a less costly spectral technique is applied. Overall, the best results for the whole classification scheme are obtained

Table 2.3 Performance results for the University of Pavia hyperspectral image (execution times in seconds)

	SVM training	SVM classification	Spect. Part. RCMG	Async. CA watershed	Majority vote	Total
CPU	0.5760 s	101.4484 s	0.1517 s	0.0186 s	0.0022 s	102.1969
GPU	3.2466 s	14.0497 s	0.0085 s	0.0010 s	0.0003 s	17.3067
Speedup	0.2×	7.2×	17.8×	18.6×	7.3×	5.9×

Table 2.4 Performance results for the Salinas Valley hyperspectral image (execution times in seconds)

	SVM Training	SVM Classification	Spect. Part. RCMG	Async. CA Watershed	Majority Vote	Total
CPU	1.5559 s	112.2305 s	0.1959 s	0.0963 s	0.0023 s	114.0809 s
GPU	11.0552 s	47.7323 s	0.0092 s	0.0035 s	0.0001 s	58.8006 s
Speedup	0.1×	2.3×	21.3×	27.5×	23.0×	1.9×

for Pavia with a speedup of 5.9×. In the next sections we will explain in detail the results for each stage.

2.5.1 SVM Spectral Classification

The standard two-class SVM spectral classification has two phases: training and classification. The training phase builds a model which is used to predict if new samples belong to one category or another in the second phase of classification, which is the most time consuming one as it can be observed in Tables 2.3 and 2.4. The percentage of time corresponding to classification is the same for both images, 81.2 % of the time in GPU required for the whole classification process.

When more than two classes are present the classification must be multiclass and different strategies can be applied in order to solve it. Hsu [46] found that the One-Against-One (OAO) method is more suitable for practical use than the One-Against-All (OAA), mainly because the total training time is shorter. In this work GPUSVM with the OAO method is used to classify the hyperspectral images. The kernel function for the SVM is a Gaussian Radial Basis Function (RBF) [28]. The number of classes considered for the classification was taken from the ground truth, with nine classes for the first image and sixteen for the second one.

First, the SVM was trained with the same values of C, γ, and number of training samples for the Pavia image as in [8]: $C = 128$, $\gamma = 0.125$ and 3192 samples. In the case of the Salinas image, the values $C = 256$ and $\gamma = 0.125$ were determined by fivefold cross validation, and the number of training samples for each class was selected as 10 % of the total samples for the class.

Table 2.5 Performance results for the spatial and spectral partitioning RCMG with the Pavia and the Salinas hyperspectral images

	Spatial partitioning		Spectral partitioning	
	Pavia	Salinas	Pavia	Salinas
CPU (OpenMP)	0.1517s	0.1959s	0.1517s	0.1959s
GPU (CUDA)	0.0537s	0.0638s	0.0085s	0.0092s
Speedup	2.8×	3.1×	17.8×	21.3×

As shown in Tables 2.3 and 2.4, the speedup results are worse for the GPU in the training phase. The SVM requires a small number of training samples in this phase [3] and, therefore, the GPU performance is low because the number of samples is not enough to exploit the big number of threads that can be simultaneously available in the GPU, up to 2,048 threads per SM in the GTX680. This is not a problem, as the training phase only needs to be performed once for each type of hyperspectral image and it is responsible for only 18.8 % of the total time.

The second phase in the spectral classification, which consumes 81.2 % of the time for the Pavia and Salinas images, obtained speedups of 7.2× and 2.3× respectively. The tests are carried out as in [40], excluding the file I/O time for both, the LIBSVM and GPUSVM, but including CPU–GPU data transfer in the GPU implementation times.

2.5.2 Robust Color Morphological Gradient

The RCMG is the vectorial gradient, described in Sect. 2.3.1, that is applied to the hyperspectral image in order to reduce it to one-band. The approaches described in Sect. 2.4.1, called spatial partitioning RCMG and spectral partitioning RCMG have been developed. Different block configurations were tested and finally the spectral partitioning RCMG implementation was configured with blocks of 32 × 4 threads. For the spatial partitioning RCMG, 128 × 4 threads per block for the Pavia image and 256 × 2 threads per block for the Salinas image were considered. Each block in the spatial partitioning RCMG processes a region of 4 × 4 pixel vectors for the first image and a region of 2 × 2 pixel vectors for the second one.

Table 2.5 shows a summary of performance for the images. The best results are for the spectral partitioning RCMG with speedups of 17.8× and 21.3×. The shared memory requirements for the spectral partitioning RCMG are 5.7 KB per block, while the spatial partitioning RCMG requires up to 20.6 KB, depending on the block size. Thus, more blocks per SM are concurrently executed in the first approach which leads to a better speedup as shown in Table 2.5.

2.5.3 *Asynchronous CA–Watershed*

As explained in Sect. 2.4.2, the asynchronous CA–Watershed takes as input the 2D gradient image obtained from the RCMG calculation. This implementation presents the advantage of reusing information within each thread block, efficiently exploiting the shared and cache memories of the GPU. In addition, it achieves better results when the image has large plateaus because in this situation the labels must be propagated through large regions. In the asynchronous implementation the labels are propagated faster within a block, unlike the synchronous implementation which performs more steps to propagate them within a plateau. Thus, less inter-block synchronizations are needed [34].

The kernels were configured to work with blocks of 32×4 threads and the shared memory was maximized to 48 KB, because only 21.4 KB are required for the 16 blocks that are simultaneously active per SM.

This proposal achieves speedups of $18.6\times$ and $27.5\times$, that can be observed in Tables 2.3 and 2.4. The speedup is better for the Salinas image, as a consequence of presenting larger plateaus.

2.5.4 *Majority Vote*

The MV was projected on the GPU taking advantage of the pipeline processing explained in Sect. 2.4, and reducing the requirements of global memory, which also means less data transfer. The times shown in Tables 2.3 and 2.4 include, as it was described in Sect. 2.4.3, the step for counting the watershed regions, as well as the global memory allocation time.

The MV obtained speedups of $7.3\times$ and $23.0\times$ for the images of Pavia and Salinas. The difference in the speedups are related mainly to the number of regions because the number of blocks executed in the *winner* step is directly related to the number of watershed regions in the image. The segmentation map of Pavia has 22,678 regions which is approximately three times more than Salinas, which has 8,423 regions, that is approximately the speedup factor observed in the performance tables.

The kernels were configured to work in one dimensional blocks. Different block sizes have been tested and it was found that the best performance is achieved for blocks of 128 threads. With this configuration each SM is fully exploited with 16 blocks simultaneously active.

2.6 Conclusions

In this work a GPU projection scheme for a spectral–spatial classification of hyperspectral images was presented. The scheme efficiently exploits the memory hierarchy and the thousands of threads available in the GPU architecture. The different stages of the scheme have been concatenated with a pipeline processing that minimizes the data transfers between the CPU and the GPU and maximizes the computational throughput. Different hyperspectral data partitioning strategies and thread block arrangements were studied in order to have a larger number of blocks being concurrently executed. The spectral classification stage was carried out with SVM using the GPUSVM, a third party library. The spatial processing stages consists in the calculation of a RCMG, that reduces the dimensionality of the hyperspectral image to a two dimensional image, followed by the asynchronous calculation of a watershed transform based on cellular automata. The spectral and the spatial results are combined by a MV technique commonly used in classification of hyperspectral images.

The projection of the classification process in the GPU requires working with data blocks of different dimensionality depending on the stage of the process: 3D for RCMG, 2D for watershed and 1D for MV. For the RCMG, two different approximations of data distribution among blocks were studied: spectral and spatial partitioning. The spectral partitioning takes better advantage of the memory hierarchy of the GPU maximizing the number of active blocks per SM. For the watershed transform an asynchronous strategy based on a cellular automaton was proposed. This asynchronous approach has the advantage that it can efficiently exploit the shared memory of the GPU being up to four times faster than a synchronous implementation. Finally, the MV was designed to save global memory space and to directly operate on the output of the other two stages using pipeline processing. This way, there is no need to move new data structures to the GPU.

The speedup values for the whole classification process were 5.9× for Pavia and 1.9× for Salinas showing the efficiency of the GPU projections while maintaining the same classification quality as when it is computed on CPU. The best performance values for the RCMG, 17.8× and 21.3×, were obtained for the spectral partitioning approach, with the images of Pavia and Salinas, respectively. The asynchronous CA–Watershed reached speedups of 18.6× and 27.5×, and the MV speedups of 7.3× and 23.0×, respectively. These results show that the GPU is an adequate computing platform for on-board processing of hyperspectral information.

As the most costly part of the spectral–spatial classification process, and therefore the critical part in terms of real-time execution, was the classification stage by SVM, other spectral classification algorithms more adequate for their efficient projection on GPU should be investigated.

Acknowledgments This work was supported in part by the Ministry of Science and Innovation, Government of Spain, cofounded by the FEDER funds of European Union, under contract TIN 2010-17541, and by Xunta de Galicia, Program for Consolidation of Competitive Research Groups ref. 2010/28. Pablo acknowledges financial support from the Ministry of Science and Innovation, Government of Spain, under a MICINN-FPI grant.

References

1. Plaza, A., Benediktsson, J.A., Boardman, J.W., Brazile, J., Bruzzone, L., Camps-Valls, G., Chanussot, J., Fauvel, M., Gamba, P., Gualtieri, A., Marconcini, M., Tilton, J.C., Trianni, G.: Recent advances in techniques for hyperspectral image processing. Remote Sens. Environ. **113**, S110–S122 (2009)
2. van der Meer, F.D., van der Werff, H.M., van Ruitenbeek, F.J., Hecker, C.A., Bakker, W.H., Noomen, M.F., van der Meijde, M., Carranza, E.J.M., de Smeth, J.B., Woldai, T.: Multi- and hyperspectral geologic remote sensing: A review. Int. J. Appl. Earth Observ. Geoinform. **14**(11), 112–128 (2012)
3. Gualtieri, J.A., Cromp, R.F.: Support vector machines for hyperspectral remote sensing classification. Proc. SPIE **3584**, 221–232 (1998)
4. Jia, X., Richards, J.A., Ricken, D.E.: Remote Sensing Digital Image Analysis: An Introduction. Springer Verlag, Berlin (1999)
5. Varshney, P. K., Arora, M. K. (eds.).: Advanced Image Processing Techniques for Remotely Sensed Hyperspectral Data. Springer Verlag, Berlin (2004)
6. Fauvel, M., Chanussot, J., Benediktsson, J.A., Sveinsson, J.R.: Spectral and spatial classification of hyperspectral data using SVMs and morphological profiles. IEEE Trans. Geosci. Remote Sens. **46**(10), 3804–3814 (2008)
7. Tarabalka, Y., Benediktsson, J.A. and Chanussot, J.: Spectral-spatial classification of hyperspectral imagery based on partitional clustering techniques. Geosci. Remote Sens. IEEE Trans. **47**(8) , 2973–2987 (2009)
8. Tarabalka, Y., Chanussot, J., Benediktsson, J.A.: Segmentation and classification of hyperspectral images using watershed transformation. Pattern Recogn. **43**(7), 2367–2379 (2010)
9. Fauvel, M.: Spectral and spatial methods for the classification of urban remote sensing data, Ph.D. Dissertation, Grenoble Institute of Technology, Grenoble (2007)
10. Farag, A.A., Mohamed, R.M., El-Baz, A.: A unified framework for MAP estimation in remote sensing image segmentation. Geosci. Remote Sens. IEEE Trans. **43**(7), 1617–1634 (2005)
11. Couprie, C., Grady, L., Najman, L., Talbot, H.: Power watershed: a unifying graph-based optimization framework. Pattern Anal. Mach. Intell. IEEE Trans. **33**(7), 1384–1399 (2011)
12. Tarabalka, Y., Benediktsson, J.A. and Chanussot, J.: Classification of hyperspectral data using support vector machines and adaptive neighborhoods. In: Proceedings of 6th EARSeL SIG IS, Workshop (2009)
13. Bernard, K., Tarabalka, Y., Angulo, J., Chanussot, J., Benediktsson, J.A.: Spectral spatial classification of hyperspectral data based on a stochastic minimum spanning forest approach. Image Process. IEEE Trans. **21**(4), 2008–2021 (2012)
14. Vincent, L., Soille, P.: Watersheds in digital spaces: an eficient algorithm based on immersion simulations. IEEE Trans. Pattern Anal. Mach. Intell. **13**, 583–598 (1991)
15. Plaza, A., Plaza, J., Vegas, H.: Improving the performance of hyperspectral image and signal processing algorithms using parallel, distributed speciallized hardware-based systems. J. Signal Proces. Syst. **61**(3), 293–315 (2010)
16. González, C., Sánchez, S., Paz, A., Resano, J., Mozos, D., Plaza, A.: Use of FPGA or GPU-based architectures for remotely sensed hyperspectral image processing. Integr. VLSI J. **46**(2), 89–103 (2013)
17. Tarabalka, Y., Haavardsholm, T.V., Kåsen, I., Skauli, T.: Real-time anomaly detection in hyperspectral images using multivariate normal mixture models and GPU processing. J. Real Time Image Process. **4**(3), 287–300 (2009)
18. Heras, D. B., Argüello, F., Gómez, J. L., Becerra, J. A., Duro, R. J.: Towards real-time hyperspectral image processing, a GP-GPU implementation of target identification. In: 2011 IEEE 6th Internatioonal Conference on Intelligent Data Acquisition and Advanced Computing Systems (IDAACS), vol. 1 pp. 316–321 (2011)
19. Priego, B., Souto, D., Bellas, F., Duro, R.J.: Unsupervised segmentation of hyperspectral images through evolved cellular automata. Adv. Knowl. Based Intell. Inform. Eng. Syst. **243**, 2160–2169 (2012)

20. Quesada-Barriuso, P., Argüello, F., Heras, D.B.: Efficient segmentation of hyperspectral images on commodity GPUs. Adv. Knowl. Based Intell. Inform. Eng. Syst. **243**, 2130–2139 (2012)
21. Christophe, E., Michel, J., Inglada, J.: Remote sensing processing: from multicore to GPU. IEEE J. Sel. Topics Appl. Earth Observ. Remote Sens. **4**(3), 643–652 (2011)
22. Bernabé, S., Plaza, A., Reddy Marpu, P., Benediktsson, J.A.: A new parallel tool for classification of remotely sensed imagery. Comput. Geosci. **46**, 208–218 (2012)
23. NVIDIA Corporation: NVIDIA CUDA C Programming Guide 4.2, Santa Clara (2011)
24. NVIDIA Corporation: NVIDIA's Next Generation CUDA Compute Architecture: Kepler GK110 Whitepaper (2012)
25. NVIDIA Corporation: CUDA C Best Practices Guide (2012)
26. Gege, P., Beran, D., Mooshuber, W., Schulz, J. and Van Der Piepen, H.: System analysis and performance of the new version of the imaging spectrometer rosis, in Proceedings of the First EARSeL Workshop on Imaging Spectroscopy. University of Zurich Remote Sensing Laboratories, pp. 29–35 (1998)
27. Green, R.O., Eastwood, M.L., Sarture, C.M., Chrien, T.G., Aronsson, M., Chippendale, B.J., Faust, J.A., Pavri, B.E., Chovit, C.J., Solis, M.S., Olah, M.R., Williams, O.: Imaging spectroscopy and the airborne visible/infrared imaging spectrometer (AVIRIS). Remote Sens. Environ. **65**(3), 227–248 (1998)
28. Camps-Valls, G., Bruzzone, L.: Kernel-based methods for hyperspectral image classification. IEEE Trans. Geosci. Remote Sens. **43**(6), 1351–1362 (2005)
29. Evans, A. and Liu, X.: A morphological gradient approach to color edge detection. Image Process IEEE Trans. **15**(6), 1454–1463 (2006)
30. Roerdink, J.B.T.M., Meijster, A.: The watershed transform: definitions, algorithms and parallelization strategies. Fund. Inform. **41**(1–2), 187–228 (2000)
31. Meyer, F.: Topographic distance and watershed lines. Math. Morphol. Appl. Signal Process. **38**(1), 113–125 (1994)
32. Galilée, B., Mamalet, F., Renaudin, M., Coulon, P.Y.: Parallel asynchronous watershed algorithm-architecture. IEEE Trans. Paral. Distrib. Syst. **18**(1), 44–56 (2007)
33. Nehaniv, C.L.: Evolution in asynchronous cellular automata. In: Proceedings of the 8th International Conference on Artificial life, MIT Press, pp. 65–73 (2003)
34. Quesada-Barriuso, P., Heras, D.B. and Argüello, F.: Efficient GPU asynchronous implementation of a watershed algorithm based on cellular automata. In: Proceedings of IEEE International Symposium on Parallel and Distributed Processing with Applications, pp. 79–86 (2012)
35. Boyer, R.S., Moore, J.S.: MJRTY - a fast majority vote algorithm, In: Boyer, R. S. (ed.) Automated Reasoning 1 Automated Reasoning Series, Springer, Netherlands, pp. 105–117 (1991)
36. Santos, A., Araújo, A., Menotti, D.: Combining multiple approaches for accuracy improvement in remote sensed hyperspectral images classification.In: Workshop of Thesis and Dissertations - XXV Conference on Graphics, Patterns and Images, pp. 54–59 (2012)
37. Balasalle, J., López, M.A., Rutherford, M.J.: Optimizing memory access patterns for cellular automata on GPUs. In: Hwu W. W. (ed.) GPU Computing Gems, Jade Edition. Morgan Kaufmann Publishers Inc., San Francisco, pp. 67–75 (2012)
38. Athanasopoulos, A., Dimou, A., Mezaris, V. and Kompatsiaris, I.: GPU acceleration for support vector machines, In: Proceedings of 12th International Workshop on Image Analysis for Multimedia Interactive Services (WIAMIS 2011), Delft (2011)
39. Carpenter, A.: CUSVM: a CUDA implementation of support vector classification and regression, patternsonscreen (2009)
40. Catanzaro, B., Sundaram, N. and Keutzer, K.: Fast support vector machine training and classification on graphics processors. In: Proceedings of the 25th International Conference on Machine learning - ICML'08, pp. 104–111 (2008)
41. Herrero-López, S., Williams, J. R. and Sanchez, A.: Parallel multiclass classification using SVMs on GPUs. In: Proceedings of the 3rd Workshop on General-Purpose Computation on Graphics Processing Units, ACM, pp. 2–11 (2010)

42. Li, Q., Salman, R., Test, E., Strack, R., Kecman, V.: GPUSVM: a comprehensive CUDA based support vector machine package. Central Eur. J. Comput. Sci. **1**(4), 387–405 (2011)
43. Harris, M.: Optimizing parallel reduction in CUDA. NVIDIA Developer Technology, (2007)
44. Hawick, K.A., Leist, A., Playne, D.P.: Parallel graph component labelling with GPUs and CUDA. Parallel Comput. **36**(12), 655–678 (2010)
45. Chang, C.C., Lin, C.J.: LIBSVM: a library for support vector machines. ACM Trans. Intell. Syst. Technol. **2**(3), 27 (2011)
46. Hsu, C.-W., Lin, C.-J.: A comparison of methods for multiclass support vector machines. IEEE Trans. Neural Netw. **13**(2), 415–425 (2002)

Chapter 3
Early Smoke Detection in Outdoor Space by Spatio-Temporal Clustering Using a Single Video Camera

Margarita Favorskaya and Konstantin Levtin

Abstract Video surveillance systems are increasingly being used to monitor urban areas and the landscape. Cameras have been proven near buildings, on bridges, ships, into tunnels. One important application of video surveillance system is the early smoke detection in the outdoor space for alarm generation. A novel video-based method of smoke detection by spatio-temporal clustering involves three developing stages. The first stage connects with any motion detection within a scene. The second stage is based on a color-texture analysis of moving regions to find smoke-like regions. Considering the complex nature of smoke (semi-transparency, spectrum overlapping, randomly motion changes) these two stages are not enough for decision making about early alarm generation. The third stage is enhanced by a spatio-temporal clustering of moving regions with a turbulence parameter connecting with fractal properties of smoke. A spatio-temporal data permit to track effectively a smoke propagation in the outdoor space by using the designed real-time software. Experimental results show that the proposed set of spatial and temporal features well discriminates smoke and non-smoke regions in outdoor scenes with a complex background.

Keywords Smoke detection · Surveillance · Turbulence · Video Sequences

3.1 Introduction

One type of surveillance an outdoor environment is an early smoke and flame detection using visual sensors by Verstockt et al. [1]. Usually the appearance of smoke precedes flame therefore the robust smoke detection is a very important task for

M. Favorskaya (✉)
Siberian State Aerospace University, Krasnoyarsk 660014, Russian Federation
e-mail: favorskaya@sibsau.ru

K. Levtin
Siberian State Aerospace University, Konstantin, Levtin, Russia
e-mail: levtin@sibsau.ru

J. W. Tweedale and L. C. Jain (eds.), *Recent Advances in Knowledge-based Paradigms and Applications*, Advances in Intelligent Systems and Computing 234,
DOI: 10.1007/978-3-319-01649-8_3,

early-warning fire alarm. Sensor-based fire alarm systems detect the presence of smoke, heat, and radiation using ionization or photometry parameters only in indoor environment. However they are not suitable in the outdoor space because combustion products may be with a low density or blown away by a strong wind that fail to produce a smoke or fire alarm. At present, computer vision technologies of smoke and flame detection can find an inflammation source even at night (Gunay et al. [2]). The existing problem is a possibility of exact smoke or flame segmentation in a real-time application. This task is divided into two categories—smoke and flame detection [3–9]. Smoke detection methods remain the priority for development of fire alarm systems. Flame detection methods also provide information of the degree of danger. Flame detection methods include the spatio-temporal fluctuation data of flame contours [10, 11], motion, flicker, edge blurring and color features [12, 13], the mixture Gaussian model for temporal features extraction [14], and some other techniques.

Toreyin et al. [15] applied motion, edge blurring, and color features for smoke detection. In this research high-frequency analysis of moving pixels was conducted in the wavelet domain. Gubbi et al. [16] found some statistical features of three-level wavelet transformed images. Interesting work was presented by Ferrai et al. [17] who proposed a real-time process for stream detection in videos based on Hidden Markov Tree and introduced stream texture patterns. For stream detection Support Vector Machine (SVM) classifier was used. Yuan [18] proposed an accumulative motion model for smoke detection. To reduce false alarms, the accumulated values are used to compensate for the inaccuracy. This paper focuses a smoke detection using the proposed method which is considered a physical specialty of smoke propagation in the outdoor space. The traditional feature set was extended by the turbulence parameter which indicates the fractal nature of smoke. The moving cluster method was applied for spatio-temporal clustering of smoke similar regions under various luminance conditions within a scene.

3.2 Related Work

One may classify the existing methods of smoke detection using both static and dynamic approaches. The static approach is based on texture analysis considering rotations and illumination differences, where the dynamic approach estimates motion changes with edge extractions and color features. Often smoke does not have a clearly defined texture or color features because of its semi-transparency nature within a spectrum. It often overlaps with the spectrum of surrounding objects. Color features of smoke received from different flame sources may randomly change. Another main aspect connects with smoke modifications in a high spectrum range which depend from a flame source, smoke intensity, and chemical specialties of flame process.

The static approach compensates for the absence of motion analysis using rotations of local patterns in 2D or 3D spaces. Ojala et al. [19] introduced Local Binary Pattern (LBP) for rotation and illumination invariant. This method was improved by a quantization step in joint histogram by Guo et al. [20] who proposed a new operator called a Local Binary Pattern Variance (LBPV). The LBPs and LBPVs features are computed at each level of the three-level image pyramid with uniform, rotation-invariance, and rotation-invariance-uniform patterns. Then, all histograms of the LBPs and LBPVs pyramids are concatenated into a feature vector for smoke detection. However this feature vector contains local and global information which is not separated. Dominant binary patterns for texture classification were suggested by Liao et al. [21]. Yuan proposed to join the LBPs and LBPVs pyramids for smoke detection in videos [22]. Celik et al. [23] proposed an algorithm which combines color-temporal information of fire and background subtraction assisted with foreground object segmentation. Therefore a static approach is characterized by a low computational cost a though the smoke region segmentation provides approximate and unstable result.

The dynamic approach is a more appropriate choice because it raises the efficiency of video-based smoke detection. Here one can find various methods of motion detection, from simple (background subtraction, temporal difference of two successive frames) up to advanced realizations (optical flow) [24, 25]. The main task is to join texture and color features with motion features of such an unstable structure as smoke in the outdoor space. Motion detection is able to eliminate stationary objects in scenes. Celik et al. [23] proposed the background subtraction method to segment the foreground object and extract color temporal information of fire. Also smoke usually blurs the edges of background objects. Such features were successfully considered in research by Toreyin et al. [12]. Yuan [18] proposed subsequently to estimate only the orientation of smoke motion to reduce the computation time.

The nature of smoke is very difficult for recognition. This connects with temporal variability of smoke density, shape, color, and texture properties. During early stage, a smoke is a semi-transparency, low density, non-robust object with difficult determined edges and pixel flicker. A view of smoke depends from a variety of factors including fuel type, quantity of oxygen, and physical situation in scene. On flame appearance stage, a smoke may become an opaque, high density, dark-grey object also with difficult determined edges. A smoke turbulent phenomenon is characterized by its fractal dimension. Catrakis et al. [26] suggested fractal similar estimations of smoke turbulence in video sequences.

Smoke clustering is a final stage of smoke segmentation based on dynamic characteristics. Size variations and non-rigid edges of smoke in temporal domain help in following smoke clustering. One of more suitable strategies connects with texture estimation of smoke regions through the co-occurrence matrix [27] or the Hurst exponent by measuring the roughness of edges [28]. The fuzzy c-means clustering algorithm was proposed by Wang et al. [29] to create the Dominant Flame Color Lookup Table (DFCLT). Both dangerous flame and smoke (it is suggested that smoke has a plum shape) are determined within a single procedure. The changed frames are automatically selected from video sequence. Then the modified regions are segmented from these frames. Finally, elementary, medium, or emergency dangerous are determined and fixed by comparing current data with data from DFCLTs.

Vidal-Calleja and Agammenoni [30] proposed the algorithm to classify objects by codebook and Bag-of-Word (BoW) paradigm. This approach does not require segmentation, extraction Region-Of-Interest (ROI), or motion computation within scenes. BoW technique represents images as a collection of regions ignoring their spatial structures. It involves three steps: a feature detection, a feature description, and a codebook generation. Such strategy is based on discriminative or generative the probabilistic probabilistic Latent Semantic Analysis (pLSA), the Bayesian form, the Latent Dirichlet Allocation (LDA) models.

This chapter describes a novel method for early smoke detection by using a single video camera in the outdoor space. The methodology contains stages that include motion analysis, color-texture analysis, and spatio-temporal clustering with the estimation of smoke turbulence. Section 3.3 presents the spatio-temporal analysis of early video-based smoke detection. In Sect. 3.4, the designed "Smoke Alarm" software is described. Section 3.5 discusses the experimental results with smoke and non-smoke moving objects obtained by using test videos. Section 3.6 offers the conclusion and future efforts.

3.3 Spatio-Temporal Analysis of Smoke Detection

The previous investigations show that only a spatio-temporal analysis is able to provide efficient intelligent decisions with their following usage in surveillance software. The following discussing methods explore several calculation of enable the motion features (Sect. 3.3.1), the color-texture features (Sect. 3.3.2), the smoke turbulence (Sect. 3.3.3), and the spatio-temporal clustering (Sect. 3.3.4). The latter is based on received features. The proposed algorithm of smoke detection is presented in Sect. 3.3.5.

3.3.1 Proposed Motion Analysis

Based on the assumption that the position of a single video camera is fixed and the observed scene is almost stationary a time-averaged background model $\mathbf{I}^b = \{\mathrm{avg}(I(x,y))\}$ can be calculated for n frames without objects motion in scene, where (x,y) are coordinates of pixels and $I(x,y)$ is a brightness function. Using the cycle in differences between current frame $\mathbf{I}^c = \{I^c(x,y)\}$ and the background model $\mathbf{I}^b$ can be determined at each pixel positioning using Eq. 3.1, where T^b is an adaptively determined threshold value from n previous frames.

$$\left| I^c(x,y) - \mathrm{avg}(I(x,y)) \right| > T^b \tag{3.1}$$

If differences detect an extended value T^b this region can be labeled as one in motion. Such differences may include motion of objects, of temporal texture (such as branches and foliage movement under a wind), and noises. Nevertheless it permits to limit regions of interest in sequential frames and reduce the computational cost of the algorithm. In the task of smoke detection this is need not exactly determine the shape of any smoke. That's why a simplified method (a block-matching method) was used when the image is divided into blocks B_k with sizes $n \times n$ pixels. The center of block B_i is located in point (x,y) at moment t and in point $(x + \Delta x, y + \Delta y)$ at moment $(t+1)$. Then values of block matching $D(B_k)$ and their displacement $\Delta\,(B_k)$ for two sequential moments t and $(t+1)$ are determined as using $D(B_k)$ and $\Delta(B_k)$.

$$D\,(B_k) = \frac{1}{n^2} \sum_{(t,i,j) \in B_k} |I_{t+1}\,(x,y) - I_t\,(x + \Delta x + i, y + \Delta y + j)\ |^2 \quad (3.2)$$

$$\Delta\,(B_k) = \sqrt{(\Delta x)^2 + (\Delta y)^2} \quad (3.3)$$

The moving regions provided by Eqs. 3.2 and 3.3 include several blocks with approximate boundaries. A procedure of edge points founding was realized by Canny detector [31]. Usually edge points of smoke region do not link, and it is necessary to build closed loops by using operators of mathematical morphology. Firstly, very short edge lines (they are concern to noises) are deleted. Secondly, a dilation operator is applied to joint the nearest edge points. Then the received boundaries are blurred for more natural view. In this way, moving regions for analysis of their color and texture features are formed.

3.3.2 Color-Texture Analysis

The color-texture analysis of moving regions includes the choice of the color space and application of filters. The choice of the color space is specified by the nature of any smoke. Often smoke has not the own color visualization because of its semi-transparent. The native spectrum may be overlapping with that of surrounding objects. The more important fact is that the color components of smoke from various sources may be randomly changed. Even during a combustion action from one source, a view of smoke is modified in a wide spectrum range determining from an intensity of source, chemical specialties of combustion, the illumination of the scene. Of the common color space models, Red, Green, Blue (RGB) and Cyan, Magenta, Yellow, blacK (CMYK)-spaces don't provide suitable components for smoke detection. Hue, Saturation, Value or Hue, Saturation, Brightness can be represented as HSV or HSB-spaces contain the saturation component S which well extracts patterns including smoke regions. Eq. 3.4 provides the saturation S by transition from RGB-space to HSV-space, where MAX and MIN are maximum and minimum values of color R, G, B components.

$$S = \begin{cases} 0 & \text{if } MAX = 0 \\ 1 - \frac{MIN}{MAX} & \text{if } MAX \neq 0 \end{cases} \quad (3.4)$$

Binary filter, brightness slice filter, histogram equalization, and brightness-contrast enhancement were chosen among main filters which can be applied in the color-texture analysis of smoke detection. Statistical texture features can be computed from a grey-level histogram. The central n-order moment of random value z (which corresponds to brightness or saturation of pixels) is calculated by Eq. 3.5, where $p(z_i), i = 0, 1, 2, \ldots, Q - 1$ is a histogram; Q is a number of grey levels.

$$\mu_n(z) = \sum_{i=0}^{Q-1} \left(z_i - \sum_{i=0}^{Q-1} z_i p(z_i) \right)^n p(z_i) \tag{3.5}$$

The second moment (dispersion) is one of the most important texture features. Eqs. 3.6–3.8 calculate a relative smoothness RS, a texture homogeneity HM, and an average entropy EN on basic of dispersion.

$$RS = 1 - 1/\left(1 + \sigma^2(z)\right) \tag{3.6}$$

$$HM = \sum_{i=0}^{Q-1} p^2(z_i) \tag{3.7}$$

$$EN = \sum_{i=0}^{Q-1} p(z_i) \quad \log_2 p(z_i) \tag{3.8}$$

In spite of all advantages of color-texture methods, they have a high inaccuracy in smoke region segmentation and false event detection. Only collaboration of dynamic and texture parameters of smoke regions permit improved segmentation results.

3.3.3 Estimation of Smoke Turbulence

A turbulent flow is a phenomenon in liquid or gas environments when nonlinear fractal waves or usual linear waves are spontaneously generated without external random forces. Various turbulence models were created for many applications. The chaotic motion of smoke may be concerned to a turbulent phenomenon. On a determined height, into a smoke puff one can observe secondary air flows coil and funnels into which objects with less sizes appear. Such coils and funnels consume the energy from the ascending smoke. Other physical and chemical processes also influence smoke turbulence. These include convection, self-oscillaltion, and hysteresis effect. In this case turbulent convection occurs [32].

In previous methods of video-based smoke detection a smoke turbulent convection as a physical phenomenon did not considered. It is known a power law relationship which describes the ratio Ω^{2D} of the region perimeter to its area in 2D-space [32] (as a rough estimation of fractal dimension [33]). Eq. 3.9 indicates the expression with empirical coefficient α^{2D}, where P is a region perimeter; A is an area.

$$\Omega^{2D} = \alpha^{2D} \frac{P}{\sqrt{A}} \approx 1.35 \tag{3.9}$$

If Eq. 3.9 is executed then a turbulent region of gaseous nature exists.

3.3.4 Spatio-Temporal Clustering

The automated spatio-temporal clustering of smoke is not a trivial task. This is explained as amorphous substance of gaseous objects. A single plum can split or join the clusters of both smoke and non-smoke regions during burning process. The color of each region will change with regard to diffusion or shadow mapping. Clustering in feature space involves only smoke cluster and non-smoke cluster; otherwise this task is a degenerate problem. The approach based on moving cluster will be appropriative decision according to dynamic properties of smoke.

Given a standardized data matrix $\mathbf{DM} = \{dm_{ij}\}$, which includes objects that look like smoke $ob_i \in \mathbf{OB}, i = 1, 3, \ldots$ and their features $ft_j \in \mathbf{FT}, j = 1, 2, \ldots, K$. Any object ob_i is characterized by a life-time and may be removed from matrix **DM** after its life-time terminates. The life-time parameters are determined from motion estimations within a set of sequential frames. A set **FT** includes all obtained features in normalized form, such as a saturation *S*, a relative smoothness *RS*, a texture homogeneity *HM*, an average entropy *EN*, and a number of others parameters. By introducing a term "cluster radius" $D > 0$, and hypothetical cluster **HS** as a set of objects the initial conditions will be determined. These objects are the nearest objects to the cluster centre of gravity g_{t0} at time moment t_0 but less cluster radius D, $\mathbf{HS} = \{ob_i : d(ob_i, g_{t0})\} < D$ where $d(ob_i, g)$ is a Euclid distance between object ob_i and a centre of gravity g. In this case, objects are sub-regions of smoke flow. During the analysis of subsequent frames, the algorithm iteratively recalculates a centre of gravity g redefining a cluster near a new centre $\mathbf{HS} = \{ob_i : d(ob_i, g_{t1})\} < D$ at moment t_1. The value of radius D is determined apriory through experimentation.

When sub-regions of smoke are clustered in a feature space, the algorithm can return to a real space and join sub-regions into one global smoke region. Such dynamic clustering is based on a spatial and a temporal clustering. At present, the main approach of spatial clustering includes a concatenation of separate smoke sub-regions into a single region and an approximation of founded single region by some geometric primitives, usually a rectangular or a square. To formalize this task: it is necessary to separate a set of points **F** on maximal amount of subsets of geometric primitives $\mathbf{G}_m$ under a condition of their non-intersection. The following algorithm is proposed:

1. Step 1. Design low-level subsets on a set of points **F** which contain connected points (x_i, y_i), (x_j, y_j). These points are close situated according to criteria $(|x_i - x_j| \leq Th_x, |y_i - y_j| \leq Th_y)$ where Th_x, Th_y are threshold values.
2. Step 2. Construct high-level subsets of geometric primitives $\mathbf{G}_m$ including low-level subsets.
3. Step 3. Repeat step 2 until the minimum amount of high-level subsets $\mathbf{G}_m$ will not be achieved.

An example of spatial clustering of smoke regions can be seen in Fig. 3.1a.

A temporal clustered region set is based on the dynamic features extract from smoke plums. The main problem is a discontinuity of smoke regions in a time domain. The algorithm uses a frame-to-frame surveillance of smoke regions under a condition that a single object may include several smoke regions. The decision about a single object is based on analysis of stochastic trajectories of the nearest smoke regions. Usually such local stochastic trajectories cross and change their directions. Closed locations and trajectories, color spectrum analysis, sizes and displacements of smoke regions are the main criteria for their temporal clustering. Only stochastic global trajectories of high-level subsets $\mathbf{G}_m$ can be built. An example of spatio-temporal clustering of smoke regions is represented in Fig. 3.1b, and global motion vectors of smoke clusters are identified in Fig. 3.1c. These examples were generated by "Smoke Alarm" software according to intelligent algorithm of smoke detection (Sect. 3.3.5).

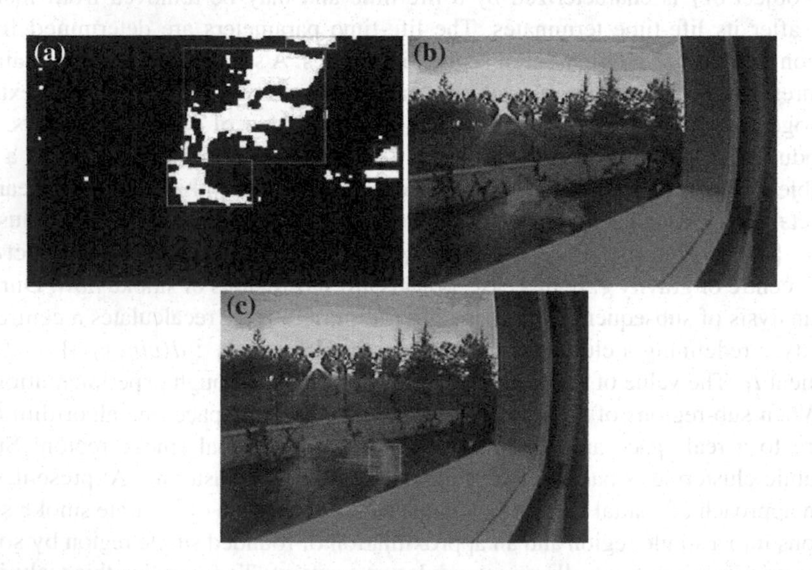

Fig. 3.1 Clustering of smoke regions: (**a**) spatial clustering; (**b**) spatio-temporal clustering; and (**c**) global motion vectors of smoke clusters

3.3.5 Intelligent Algorithm of Smoke Detection

The proposed algorithm is based on six stages:

1. A frames pre-processing.
2. A motion detection.
3. A color-texture segmentation.
4. An edges removal.
5. A morphological post-processing.
6. A spatio-temporal clustering.

The novelty of the algorithm is in the extraction of the enhanced set of features from video sequences. A pre-processing stage includes brightness and contrast normalization of frames with noise filtering. Single spikes are excluded because they are not good for following motion and turbulence estimations. A median filter uses the adaptive aperture for suppression of weak-correlated noises and low-sized details on images.

Motion detection, color-texture segmentation, and spatio-temporal clustering are discussed in Sects. 3.3.1, 3.3.2, and 3.3.4 respectively. Detection and removal of well-defined edges from sequential frames permit to find smoke regions because smoke usually blurs the objects images. The algorithm of detection and removal edges contains four steps:

1. An edges detection based on Laplacian filter.
2. An adaptive binary filtration.
3. A morphological dilatation.
4. A subtraction the processed frame from the initial frame received after a color-texture segmentation.

A morphological post-processing is used for integration of closed regions on frame. With high probability, such closed regions may be concerned to smoke regions. A morphological processing uses the adaptively choice of window slice sizes (according to regions sizes). Morphological operations of closing and opening are sequentially applied for the processing frame.

3.4 Software "Smoke Alarm"

In the designed "Smoke Alarm" software, the spatio-temporal clustering of smoke regions is realized by a differentiated color labeling. Regions are clustered using non-directed motion of sub-regions and by founding their motion vectors.

General activity diagram is represented on Fig. 3.2. The pre-processing stage includes frames partitioning from video stream, brightness and contrast normalization of frames, and also noise filtering. Motion detection is based on block-matching method. Color and texture features of moving regions are the main estimations for

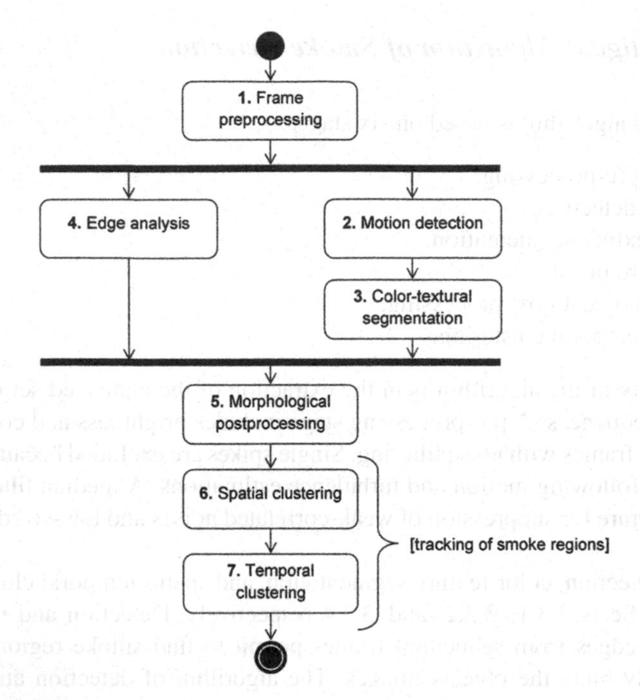

Fig. 3.2 General activity diagram of "Smoke Alarm" software

spatial clustering. The additional estimations are calculated by using edge information (fractal dimension, wavelet coefficients, and flicker effect). Such additional estimations are applied in the basic of temporal clustering. Morphological processing is the necessary stage after spatial and temporal feature extraction because of usual "garbage" after any digital processing of continuous 2D video signal.

The "Smoke Alarm" software has three functional modules: Correction Module, Detection Module, and Comparison Module that generates results.

The Correction Module permits a manual tuning of all parameters of video processing and a choice of adaptive modes. All parameters are represented as parameters of common tuning, color analysis, texture analysis, motion analysis, and spatio-temporal clustering. Input data enters from a user as numeric values or by options' selections. Output data passes to the Detection Module.

The Detection Module is the central core which executes the main functions and joins all other Modules. According to frames from video sequence and determined tunings, this Module realizes a search and a following labeling of smoke regions. Some parameters may be calculated automatically in the case of a current video stream. Processed images are displayed in real-time; they may be saved on Hard Disk Drive (HDD) in bitmap-format for subsequent analysis in the Comparison Module.

The Comparison Module calculates some parameters between current and pattern smoke images in unreal-time. There are true detected elements, undetected elements, false detected elements, and sizes of detected regions.[1] Output data are displayed on screen and stored in a session journal during a program execution.

The main screen of "Smoke Alarm" software is represented in Fig. 3.3. It includes several tabs such as "Processing Video", "Morphology/Edge Method", "Textural/Block Method", and "Marked Area Comparison" with multiple tuning parameters. Designed software includes following procedures: transitions between color spaces, texture analysis, procedure of motion estimation, object clustering, fractal estimation, wavelet transformation, flicker estimation, tuning maintenance, 2D Cleaner filtering, median filtering, sorting of point array, and additional procedures and functions.

The working of expert is simplified by using of lasso or polygonal lasso programming tools. Example of polygonal lasso is seen in Fig. 3.4.

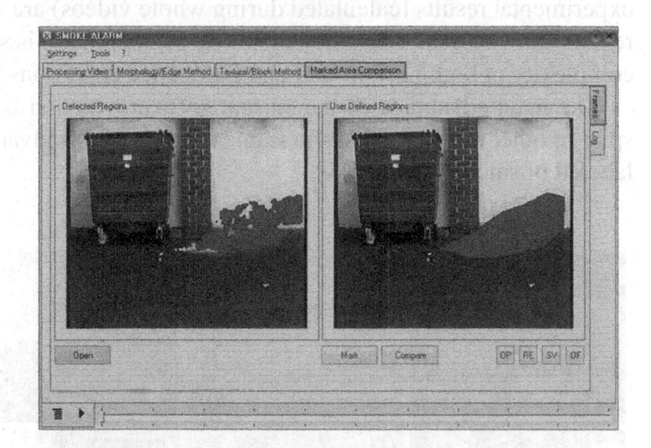

Fig. 3.3 Main screen of "Smoke Alarm" software

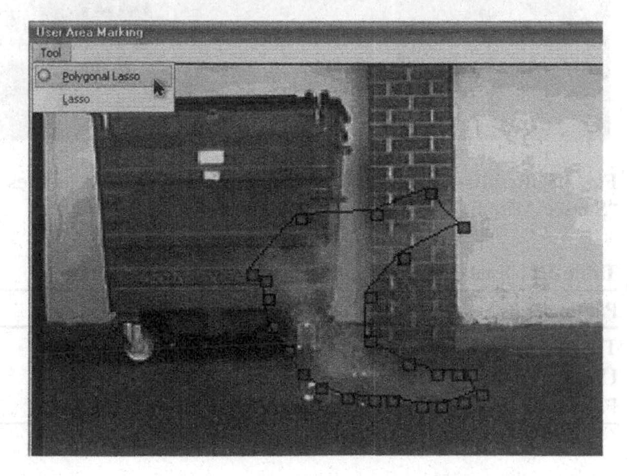

Fig. 3.4 Example of polygonal lasso

[1] Pattern smoke image is an image where smoke regions are labeled by the expert.

The "Smoke Alarm" software is used during each experimental to process video-based smoke detection in the outdoor space which is realized in the environment of Rapid Application Design "Borland Delphi 7". A free component DSPACK was used to work with video stream by utility *DirectShow* and a set of free components *AlphaControls* for the interface design.

3.5 Experimental Results

The efficiency of proposed methods and algorithms was determined by processing of some video sequences from a test Dataset.[2] Automatic and manual results of smoke segmentation in test frames of video sequences are represented in Fig. 3.5. Total experimental results (calculated during whole videos) are situated in Table 3.1. The results show that the efficiency of smoke segmentation is strongly dependant of the complexity of both background and luminance conditions.

The main advantage of the current software version is true smoke detection in spite of other moving objects in frames. The main disadvantage manifests as poorly labeled of smoke regions.

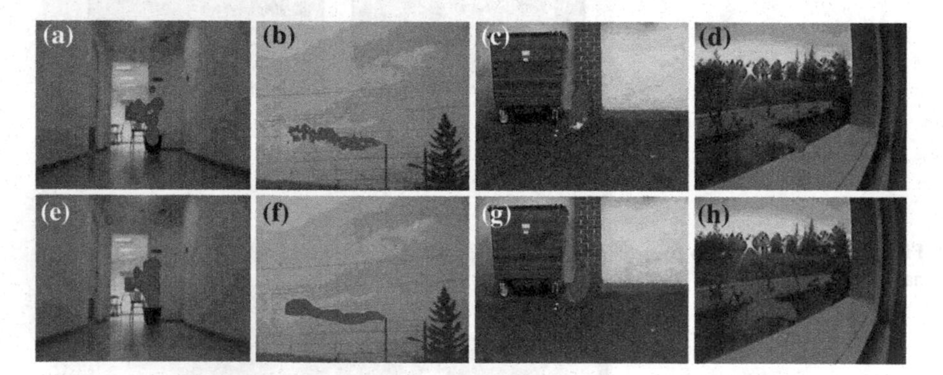

Fig. 3.5 Results of smoke segmentation for 4 test video sequences: **(a–d)** segmentation results by "Smoke Alarm" software; **(e–h)** manual segmentation results

Table 3.1 Total experimental results of smoke segmentation

Parameter (%)	Test 1	Test 2	Test 3	Test 4
True detected elements	63.45	60.33	79.03	53.53
Undetected elements	36.55	39.67	20.97	46.47
False detected elements	10.52	23.17	30.99	20.02

[2] *D*ataset, http://signal.ee.bilkent.edu.tr/VisiFire/Demo/SmokeClips

3.6 Conclusion

At present, a video-based detection is an effective and inexpensive method of identifying smoke and fire alarms in the outdoor space. This novel method is based on spatio-temporal clustering of smoke region calculations and builds trajectories, motion vectors of smoke streams from simultaneous active sources. Experimental software called "Smoke Alarm" produces good results for smoke and non-smoke object segmentation in outdoor scenes (with a complex background). This investigations will developed in fractal and frequency domains by using a statistical self-similarity and a wavelet analysis for the purpose of smoke segmentation improvement. Another way to achieve a high efficiency is in a combined data from visual and temperature sensors. Visual methods are necessary for early alarms in dangerous situations; temperature control is used for elimination of false events in complex situations when not only a smoke but a gas stream of another nature propagates in the outdoor space.

References

1. Verstockt, S., Merci, B., Lambert, P., van de Walle, R., Sette, B.: State of the art in vision-based and smoke detection. In: Proceedings of the 14th International Conference on Auto-matic Fire Detection, vol. 2, pp. 285–292 (2009)
2. Gunay, O., Tasdemir, K., Toreyin, U., Cetin, A.E.: Video based wildfire detection at night. Fire Saf. J. **44**, 860–868 (2009)
3. Han, D., Lee, B.: Flame and smoke detection method for early real-time detection of a tunnel fire. Fire Saf. J. **44**, 951–961 (2009)
4. Ho, C.-C.: Machine vision-based real-time early flame and smoke detection. Meas. Sci. Technol. **20**(4), 450–502 (2009)
5. Ko, B.C., Cheong, K.-H., Nam, J.-Y.: Fire detection based on vision sensor and support vector machines. Fire Saf. J. **44**(3), 322–329 (2009)
6. Qi, X., Ebert, J.: A computer vision-based method for fire detection in color videos. Int. J. Imaging **2**(S09), 22–34 (2009)
7. Celik, T., Demirel, H.: Fire detection in video sequences using a generic color model. Fire Saf. J. **44**, 147–158 (2009)
8. Chen, J., He, Y., Wang, J.: Multi-feature fusion based fast video flame detection. Build. Environ. **45**, 1113–1122 (2010)
9. Habiboglu, Y.H., Gunay, O., Cetin A.E.: Real-time wildfire detection using correlation descriptors. In: 19th European Signal Processing Conference, EUSIPCO 2011, pp. 894–898 (2011)
10. Yamagishi H., Yamaguchi J.: A contour fluctuation data processing method for fire flame detection using a color camera. In: IEEE 26th Annual Conference on IECON of the Industrial Electronics Society, vol. 2, pp. 824–829 (2000)
11. Piccinini, P., Calderara, S., Cucchiara, R.: Reliable smoke detection in the domains of image energy and color. In: Proceedings of the 15th IEEE Conference on Image Processing, pp. 1376–1379 (2008)
12. Toreyin, B.U., Dedeoglu, Y., Cetin, A.E.: Computer vision based method for real-time fire and flame detection. Pattern Recogn. Lett. **27**(1), 49–58 (2006)
13. Yasmin, R.: Detection of smoke propagation direction using color video sequences. Int. J. Soft Comput. **4**(1), 45–48 (2009)

14. Yuan, F.N., Liao, G.X., Fan, W.C., Zhou, H.Q.: Vision based fire detection using mixture Gaussian model. In: Proceedings of the 8th International Symposium on Fire Safety Science, vol. 8, pp. 1575–1583 (2005)
15. Toreyin, B.U., Dedeoglu, A.Y., Cetin, E.: Wavelet based real-time smoke detection in video. In: Proceedings of the 13th European Signal Processing Conference EUSIPCO, pp. 4–8 (2005)
16. Gubbi, J., Marusic, S., Palaniswami, M.: Smoke detection in video using wavelets and support vector machines. Fire Saf. J. **44**(8), 1110–1115 (2009)
17. Ferrari, R.J., Zhang, H., Kube, C.R.: Real-time detection of steam in video images. Pattern Recogn. **40**(3), 1148–1159 (2007)
18. Yuan, F.: A fast accumulative motion orientation model based on integral image for video smoke detection. Pattern Recogn. Lett. **29**(7), 925–932 (2008)
19. Ojala, T., Pietikainen, M., Maenpaa, T.T.: Multiresolution gray-scale and rotation invariant texture classification with local binary pattern. IEEE Transactions on Pattern Analysis and Machine Intelligence, **24**(7), 971–987 (2002)
20. Guo, Z.H., Zhang, L., Zhang, D.: Rotation invariant texture classification using LBP variance (LBPV) with global matching. Pattern Recogn. **43**(3), 706–719 (2009)
21. Liao, S., Law, M.W.K., Chung, C.S.: Dominant local binary patterns for texture classification. IEEE Trans. Image Process. **18**(5), 1107–1118 (2009)
22. Yuan, F.: Video-based smoke detection with histogram sequence of LBP and LBPV pyra-mids. Fire Saf. J. **46**, 132–139 (2011)
23. Celik, T., Demirel, H., Ozkaramanli, H., Uyguroglu, M.: Fire detection using statistical color model in video sequences. J. Vis. Commun. Image Represent. **18**(2), 176–185 (2007)
24. Yu, C., Zhang, Y., Fang, J., Wang, J.: Video smoke recognition based on optical flow. In: Proceedings of the 2th International Conference on Advanced Computer Control, vol. 2, pp. 16–21 (2010)
25. Favorskaya, M.: Motion estimation for object analysis and detection in videos. In: Kountchev, R., Nakamatsu, K. (eds.) Advances in reasoning-based image processing, analysis and intelligent systems: Conventional and intelligent paradigms, pp. 211–253. Springer-Verlag, Berlin Heidelberg (2012)
26. Catrakis, H.J., Dimotakis, P.E.: Shape Complexity in Turbulence. Phys. Rev. Lett. **80**(5), 968–971 (1998)
27. Maruta, H., Nakamura, A., Kurokawa, F.: A novel smoke detection method using support vector machine. In: IEEE TENCON, pp. 210–215 (2010)
28. Maruta, H., Nakamura, A., Yamamichi, T., Kurokawa, F.: Image based smoke detection with local Hurst exponent. In: IEEE TENCON, pp. 4653–4656 (2010)
29. Wang, S.J., Jeng, D.L., Tsai, M.T.: Early fire detection method in video for vessels. J. Syst. Softw. **82**(4), 656–667 (2009)
30. Vidal-Calleja, T.A., Agammenoni, G.: Integrated probabilistic generative model for detecting smoke on visual images. In: IEEE International Conference on Robotics and Automation River Centre, pp. 2183–2188. ACFR, Saint Paul, Minnesota (2012)
31. Canny, J.: A computational approach to edge detection. IEEE Trans. Pattern Anal. Mach. Intell. (PAMI) **8**(6), (1986) 679–698
32. Catrakis, H.J., Aguirre, R.C., Ruiz-Plancarte, J., Thayne, R.D.: Shape complexity of whole-field three-dimensional space-time fluid interfaces in turbulence. Phys. Fluids **14**(11), 3891–3898 (2002)
33. Favorskaya, M.N., Petukhov, N.Y.: Recognition of natural objects on air photographs using neural networks. J. Opt. Instrum. Data Process. **47**(3), 233–238 (2011)

Chapter 4
Using Evolved Artificial Neural Networks for Providing an Emergent Segmentation with an Active Net Model

Jorge Novo, Cristina V. Sierra, José Santos and Manuel G. Penedo

Abstract A novel segmentation method using deformable models for medical image segmentation was developed. As deformable model, Topological Active Nets (TAN) were used, model which integrates features of region-based and boundary-based segmentation techniques. The model deformation through time is controlled by an Artificial Neural Network (ANN) that learns how to move the nodes of the model based on their energy surrounding. The ANN is applied to each of the nodes and in different temporal steps until the final segmentation is reached. The ANN training is obtained by simulated evolution, using Differential Evolution (DE) to automatically obtain the ANN that provides the emergent segmentation. The proposed methodology was adapted and tested in three different medical domains, that is, Computed Tomography (CT), Cone Beam Computed Tomography (CBCT) and eye fundus images to demonstrate the potential of the segmentation technique.

Keywords Topological Active Nets · Differential Evolution · Artificial Neural Networks · Medical Imaging

J. Novo (✉) · C. V. Sierra · J. Santos · M. G. Penedo
Computer Science Department, University of A Coruña, A Coruña, Spain
e-mail: jnovo@udc.es

C. V. Sierra
e-mail: cristina.delavega@udc.es

J. Santos
e-mail: santos@udc.es

M. G. Penedo
e-mail: mgpenedo@udc.es

J. W. Tweedale and L. C. Jain (eds.), *Recent Advances in Knowledge-based Paradigms and Applications*, Advances in Intelligent Systems and Computing 234,
DOI: 10.1007/978-3-319-01649-8_4, © Springer International Publishing Switzerland 2014

4.1 Introduction and Previous Work

The active nets model for image segmentation was proposed by Tsumiyama and Yamamoto [1] as a variant of deformable models [2] that integrates features of region–based and boundary–based segmentation techniques. To this end, active nets distinguish two kinds of nodes: internal nodes, related to the region–based information, and external nodes, related to the boundary–based information. The former model the inner topology of the objects whereas the latter fit the edges of the objects. The Topological Active Net (TAN) [3] model was developed as an extension of the original active net model [1]. It solves some intrinsic problems to the deformable models such as the initialization problem. It also has a dynamic behavior that allows topological local changes in order to perform accurate adjustments and find all the objects of interest in the scene. The model deformation is controlled by energy functions in such a way that the mesh energy has a minimum when the model is over the objects of the scene. This way, the segmentation process turns into a minimization task.

The energy minimization of a given deformable model has been faced with different minimization techniques. One of the simplest methods is the greedy strategy, proposed by Williams and Shah [4]. The main idea implies the local modification of the model in a way the energy of the model is progressively reduced. The segmentation finishes when no further modification implies a reduction in terms of energy. As the main advantages, this method is fast and direct, providing the final segmentations with low computation requirements. However, as a local minimization method, it is also sensitive to possible noise or complications in the images. This method was used as a first approximation to the energy minimization of the TANs [3].

As the local greedy strategy presented important drawbacks, especially regarding the segmentation with complex and noisy images, different global search methods based on evolutionary computation were proposed. Thus, Ibáñez et al. [5] designed a global search method using Genetic Algorithm (GAs). As a global search technique, this method provided better results working under different complications in the image, like noise or fuzzy and complex boundaries, situations quite common working under real conditions. This method requires definition of ad-hoc genetic operators and different evolutionary phases with different aims in order to obtain desirable segmentations. The method was applied to practical domains. In particular, it was used for the Optic Disc (OD) detection in eye fundus images [6]. This strategy was also extended for 3D segmentations, for the Topological Active Volume (TAV) model optimization, adapting the genetic characteristics and introducing other requirements [7].

However, this approach presented an important drawback, that is the complexity. The segmentation process needed large times and computation requirements to reach the desired results. As an improvement of the GA approach, another evolutionary optimization technique was used by Novo et al. [8]. This approach, based on Differential Evolution (DE), allowed a simplification of the previous method and also speeded up the segmentation process, obtaining the final results in less generations (implying less computation time). As an alternative paradigm, a segmentation strategy using Multiobjective Optimization Evolutionary Algorithm (MOEAs) was implemented. These algorithms consider the optimization of several objectives in parallel. They usually work with conflicting objectives, while trying to identify a set of optimal trade-off solutions labelled the *Pareto Set*. A methodology based on one of the best established algorithms of this type, the Strength Pareto Evolutionary Algorithm 2 (SPEA2) [9], was previously defined, methodology that was adapted to this specific domain [10, 11].

There is very little work regarding emerging systems and deformable models for image segmentation, specially regarding medical imaging. McInerney et al. [12] used "deformable organisms" to automate the segmentation of medical images. Their artificial organisms possessed deformable bodies with distributed sensors, while their behaviors consisted of movements and alterations of predefined body shapes (defined in accordance with the image object to segment). The authors demonstrated the method with several prototype deformable organisms based on a multiscale axisymmetric body morphology, including a "corpus callosum worm" to segment and label the corpus callosum in 2D mid-sagittal Magnetic Resonance (MR) brain images.

In this paper, DE [13, 14] was used to train an Artificial Neural Network (ANN) that works as a "segmentation operator" that knows how to move each TAN node in order to reach the final segmentations. Section 4.2 details the main characteristics of the method. It includes the basis of the TAN, deformable model used to achieve the segmentations (Sect. 4.2.1), the details of the ANN designed (Sect. 4.2.2) and the optimization of the ANN parameters using the DE method (Sect. 4.2.3). Section 4.3 depicts the application of the method in three different medical domains. In particular, different Computed Tomography (CT), Cone Beam Computed Tomography (CBCT) and retinal images are used to show the results and capabilities of the approach. Finally, Section 4.4 expounds the conclusions of the work.

4.2 Methods

4.2.1 Topological Active Nets

A TAN is a discrete implementation of an elastic $n-$dimensional mesh with interrelated nodes [3]. The model has two kinds of nodes: internal and external. Each kind of node represents different features of the objects: the external nodes fit their edges whereas the internal nodes model their internal topology. Figure 4.1 shows a TAN mesh that contains both types of nodes.

A TAN is defined parametrically as $v(r, s) = (x(r, s), y(r, s))$ where $(r, s) \in ([0, 1] \times [0, 1])$. As other deformable models, the mesh deformations are controlled by an energy function defined as:

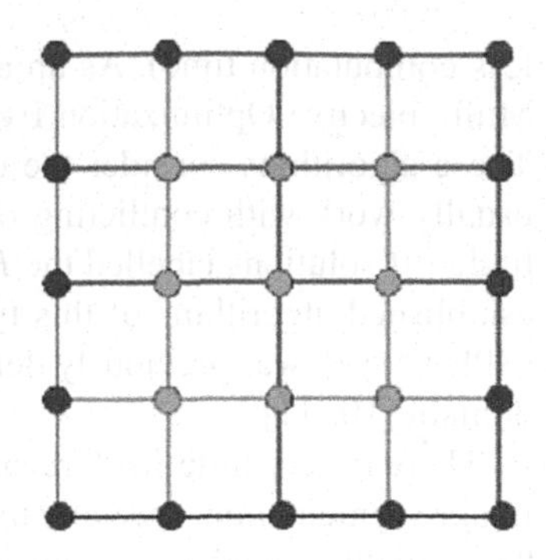

Fig. 4.1 *A*5 × 5 TAN mesh. The external nodes are on the boundaries (*printed in blue*), whereas the internal nodes are inside the mesh (printed in green)

$$E(v(r,s)) = \int_0^1 \int_0^1 (E_{int}(v(r,s)) + E_{ext}(v(r,s)))drds \quad (4.1)$$

where E_{int} and E_{ext} are the internal and the external energy of the TAN, respectively. The internal energy controls the shape and the structure of the net whereas the external energy represents the external forces which govern the adjustment process. These energies are composed of several terms and in all the cases the aim is their minimization.

Internal energy terms. The internal energy depends on first and second order derivatives which control contraction and bending, respectively. The internal energy term is defined through the following equation for each node:

$$\begin{aligned} E_{int}(v(r,s)) = {} & \alpha\left(|v_r(r,s)|^2 + |v_s(r,s)|^2\right) \\ & +\beta\left(|v_{rr}(r,s)|^2 + |v_{rs}(r,s)|^2 + |v_{ss}(r,s)|^2\right) \end{aligned} \quad (4.2)$$

where the subscripts represent partial derivatives, and α and β are coefficients that control the first and second order smoothness of the net. In order to calculate the energy, the parameter domain $[0, 1] \times [0, 1]$ is discretized as a regular grid defined by the internode spacing (k, l) and the first and second derivatives are estimated using the finite differences technique. On one hand, the first derivatives are computed using the following equations to avoid the central differences:

$$|v_r(r,s)|^2 = \frac{||d_r^+(r,s)||^2 + ||d_r^-(r,s)||^2}{2} \qquad |v_s(r,s)|^2 = \frac{||d_s^+(r,s)||^2 + ||d_r^-(r,s)||^2}{2} \quad (4.3)$$

where d^+ and d^- are the forward and backward differences respectively, which are computed as follows:

$$\begin{aligned} d_r^+(r,s) &= \frac{v(r+k,s) - v(r,s)}{k} & d_r^-(r,s) &= \frac{v(r,s) - v(r-k,s)}{k} \\ d_s^+(r,s) &= \frac{v(r,s+l) - v(r,s)}{l} & d_s^-(r,s) &= \frac{v(r,s) - v(r,s-l)}{l} \end{aligned} \quad (4.4)$$

On the other hand, the second derivatives are estimated by:

$$v_{rr} = \frac{v(r-k,s)-2v(r,s)+v(r+k,s)}{k^2} v_{ss} = \frac{v(r,s-l)-2v(r,s)+v(r,s+l)}{l^2}$$
$$v_{rs}(r,s) = \frac{v(r-k,s)-v(r-k,s+l)-v(r,s)+v(r,s+l)}{kl} \tag{4.5}$$

External energy terms. The external energy represents the features of the scene that guide the adjustment process:

$$E_{ext}(v(r,s)) = \omega f[I(v(r,s))] + \frac{\rho}{|\aleph(r,s)|} \sum_{p \in \aleph(r,s)} \frac{1}{\|v(r,s) - v(p)\|} f[I(v(p))] \tag{4.6}$$

where ω and ρ are weights, $I(v(r,s))$ is the intensity of the original image in the position $v(r,s)$, $\aleph(r,s)$ is the neighborhood of the node (r,s) and f is a function, which is different for both types of nodes since the external nodes must fit the edges whereas the internal nodes model the inner features of the objects.

If the objects to detect are bright and the background is dark, the energy of an internal node will be minimum when it is on a point with a high gray level. Also, the energy of an external node will be minimum when it is on a discontinuity and on a dark point outside the object. Given these circumstances, the function f is defined as:

$$f[I(v(r,s))] = \begin{cases} IO_i(v(r,s)) + \tau IOD_i(v(r,s)) & \text{for internal nodes} \\ IO_e(v(r,s)) + \tau IOD_e(v(r,s)) + \xi(G_{max} - G(v(r,s))) + \delta GD(v(r,s)) & \text{for external nodes} \end{cases} \tag{4.7}$$

where τ, ξ and δ are weighting terms, G_{max} and $G(v(r,s))$ are the maximum gradient and the gradient of the input image in node position $v(r,s)$, $I(v(r,s))$ is the intensity of the input image in node position $v(r,s)$, IO is a term called "In-Out" and IOD a term called "distance In-Out", and $GD(v(r,s))$ is a gradient distance term. The IO term minimizes the energy of individuals with the external nodes in background intensity values and the internal nodes in object intensity values meanwhile the terms IOD act as a gradient: for the internal nodes (IOD_i) its value minimizes towards brighter values of the image, whereas for the external nodes its value (IOD_e) is minimized towards low values (background).

The adjustment process consists of minimizing these energy functions, considering a global energy as the sum of the different energy terms, weighted with the different exposed parameters, as used in the optimizations with a greedy algorithm [3] or with an evolutionary approach [5, 8].

4.2.2 *Artificial Neural Networks for Topological Active Nets Deformation*

Our proposal is a segmentation technique that uses ANNs to perform the optimization of the TANs. A classical multilayer perceptron model was used, trained to know how the TAN nodes have to be iteratively moved and reach the desired segmentations.

The main purpose of the ANNs consist of providing, for a given node, the most suitable movement that implies an energy minimization of the whole TAN structure. This is not the same as the greedy algorithm, which determines the minimization for each node movement. All the characteristics of the network were designed to obtain this behavior, and are the following:

Input: The ANN is applied iteratively to each of the TAN nodes. The network has as input the four hypothetical energy values that would take the mesh if the given node was moved in the four cardinal directions. In addition, these values are normalized with respect to the energy in the present position, given the high values that energy normally takes, following the formula:

$$\Delta E_i = (E_i - E_c)/E_c \tag{4.8}$$

where E_i is the given hypothetical energy in the new position and E_c is the energy with the TAN node in the present location.

Hidden layers: One single hidden layer composed by a different number of nodes.

Output: The network provides the movement that has to be done in each axis for the given TAN node. So, it has two output nodes that specify the shift in both directions of the current position. The sigmoid transfer function was used for all the nodes.

These characteristics are summarized in Fig. 4.2. In this case, we obtain the values of the hypothetical energies that would be taken if we move the central node in the x and y axes, represented by the E_{x-}, E_{x+}, E_{y-} and E_{y+} values. The increments of energy with respect to the current position (Eq. 4.8) are introduced as the input values in the corresponding ANN, that produces, in this example, a horizontal displacement for the given ANN node. This movement, defined by the network outputs, is restricted

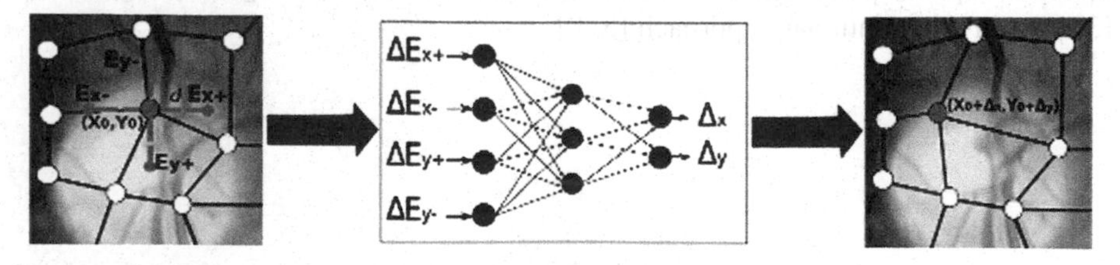

Fig. 4.2 Diagram of the use of the ANN for controlling the movement or shift of each of the TAN nodes

to a small interval of pixels around the current position, typically between one and five pixels in both axes and directions.

Once the ANN is correctly trained (with the evolutionary algorithm, as explained next), it can be used as a "segmentation operator" that progressively moves the entire set of TAN nodes until, after a given number of steps, the TAN reaches the desired segmentation. In this process, the ANN is applied to each of the nodes sequentially. Such a temporal "step" is the application of the ANN to all the nodes of the TAN. Figure 4.3 shows a couple of examples of the evolution of a segmentation process, using an artificial image and a CT image. The TAN was initially established in

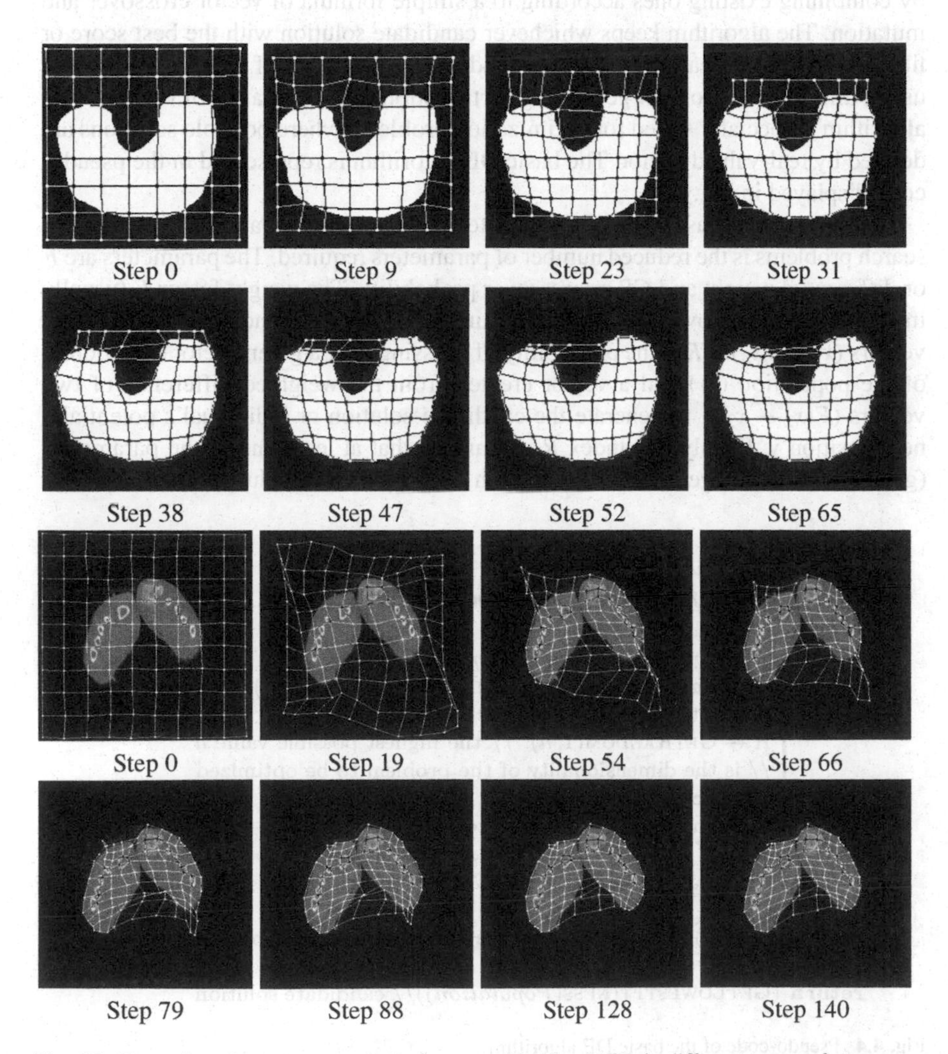

Fig. 4.3 Examples of emergent segmentations provided by ANNs in different temporal steps

the limits of the image and all the nodes were progressively moved until a correct segmentation was reached.

4.2.3 Differential Evolution for the optimization of the Artificial Network

DE [13, 14] is a population-based search method. DE creates new candidate solutions by combining existing ones according to a simple formula of vector crossover and mutation. The algorithm keeps whichever candidate solution with the best score or fitness on the optimization problem at hand. The central idea of the algorithm is the use of difference vectors for generating perturbations in a population of vectors. This algorithm is specially suited for optimization problems where possible solutions are defined by real-valued vector. The basic DE algorithm is represented in the pseudo-code displayed in Fig. 4.4.

One of the main reasons why DE is an interesting method in many optimization and search problems is the reduced number of parameters required. The parameters are F or differential weight and CR or crossover probability. The weight factor F (usually in [0, 2]) is applied over the vector resulting form the difference between pairs of vectors (x_2 and x_3). CR is the probability of crossing over a given vector (individual) of the population (x_1) and a vector created from the weighted difference of two vectors ($F(x_2 - x_3)$), to generate the candidate solution or individual's potentially new position y. Finally, the index R guarantees that at least one of the parameters (genes) will be changed in such generation of the candidate solution.

```
for each Individual ∈ Population
  do {Individual ← InitializeRandomPositions()
repeat
  for each Individual x ∈ Population
      { x1, x2, x3 ← GetRandomIndividual(Population)
      { // must be distinct from each other and x
      { R ← GetRandom(1, n)  // the highest possible value n
      { // is the dimensionality of the problem to be optimized
   do { for each i ∈ 1 : n
      { // Compute individual's potentially new position y = [y1, ..., yn]
      {        { ri ← GetRandom(0, 1)// uniformly in open range (0,1)
      {     do { if ((i = R) || (ri < CR))  yi = x1i + F(x2i − x3i)
      {        {   else yi = xi
      { if (f(y) < f(x))  x = y// replace x with y in Population
until TerminationCriterion()
return (GetLowestFitness(Population))// candidate solution
```

Fig. 4.4 Pseudo-code of the basic DE algorithm

When compared with the classical evolutionary algorithms such as GAs, DE has a clear advantage. The main problem of the GAmethodology is the need of tuning of a series of parameters: probabilities of different genetic operators such as crossover or mutation, decision of the selection operator (tournament, roulette.), tournament size... Hence, in a standard GA it is difficult to control the balance between exploration and exploitation. On the contrary, DE reduces the parameters tuning and provides an automatic balance in the search. As Feoktistov [15] indicates, the fundamental idea of the algorithm is to adapt the step length ($F(x_2 - x_3)$) intrinsically along the evolutionary process. At the beginning of generations the step length is large, because individuals are far away from each other. As the evolution goes on, the population converges and the step length becomes smaller and smaller.

In this application, a single ANN was used to learn the movements that have to be done by the internal and the external nodes. In the evolutionary population, each individual encodes the ANN. The genotypes code all the weights of the connections between the different nodes of the ANN. The weights were encoded in the genotypes in the range $[-1, 1]$, and decoded to be restricted in an interval $[-MAX\_VALUE, MAX\_VALUE]$. In the current ANN used, the interval $[-1, 1]$ was enough to determine output values in the whole range of the transfer functions of the nodes.

An initial TAN with a square topology with interrelated nodes was used, which covers the whole image (Fig. 4.3, step 0), and the ANN was applied a fixed number of steps. Each step consists of the modification produced by the ANN for each of the nodes of the TAN. Finally, the fitness associated to each individual or encoded ANN is the energy that has the final emergent segmentation provided by an encoded ANN.

In addition, the usual implementation of DE chooses the base vector x_1 randomly or as the individual with the best fitness found up to the moment (x_{best}). To avoid the high selective pressure of the latter, the usual strategy is to interchange the two possibilities across generations. Instead of this, a tournament was used to pick the vector x_1, which allows us to easily establish the selective pressure by means of the tournament size.

4.3 Results

The methodology was tested in three medical domains: CT images, CBCT images and retinal images. Different representative images were selected to show the capabilities and advantages of the proposed method.

In the evolutionary DE optimization of the ANNs , all the processes used a population of 1000 individuals. The tournament size to select the base individual x_1 in the DE runs was 5 % of the population. Fixed values were used for the *CR* parameter (0.9) and for the *F* parameter (0.9), values that empirically provided the best results. In the calculation of the fitness of the individual, a fixed number of steps were applied between 200 and 300, depending on the complexity and the resolution of the image.

Table 4.1 TAN parameter sets used in the segmentation processes of the examples

Figures	Size	α	β	ω	ρ	ξ	δ	τ
4.5, 4.6	8 × 8	4.5	0.8	10.0	2.0	7.0	20.0	40.0
4.7, 4.8	10 × 10	0.1	2.0	10.0	4.0	5.0	10.0	50.0
4.9, 4.10, 4.11	8 × 8	1.5	0.5	10.0	0.0	0.0	0.0	30.0

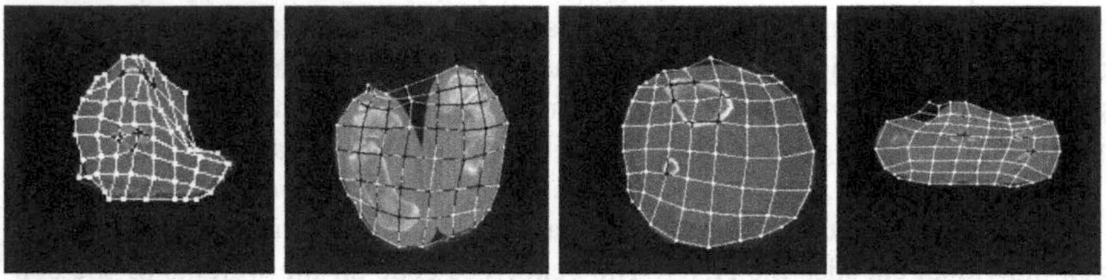

Fig. 4.5 Results obtained with the best evolved ANN and the training set of real CT images

Table 4.1 includes the energy TAN parameters used in the segmentation examples. Those were experimentally set as the ones in which the corresponding ANN gave the best results for each training.

4.3.1 Segmentation of CT images

Firstly, the methodology was tested with CT images with different characteristics. The ANNs were "trained" or evolved with a set of medical CT images, and after that, the method was tested with a different dataset. When a training set of images for evolving the ANN is used, the same ANN was applied the fixed number of steps to the different images and beginning with the initial TAN that covers the whole image. The fitness is defined as the sum of the final fitness (energy) provided by the final segmentations in each of the individual images. The number of generations in the DE evolutions was between 100 and 150 to obtain the optimized TAN.

A set of images that included objects with different shapes and levels of complexity was selected. All these CT images presented some noise surrounding the object, noise that was introduced by the capture machines when the medical CT images were obtained.

Figure 4.5 shows the segmentations obtained after the training process with the training dataset of four different CT images. The CT images correspond to a CT image of the head, the feet, the knee and a CT image at the level of the shoulders.

The images used in the testing correspond to CT images of the same close areas of the corresponding images in the training set, but with slightly different shapes and with deeper concavities. Figure 4.6 details the final segmentations obtained with the best trained ANN. In both cases, training and test dataset, the evolved ANN

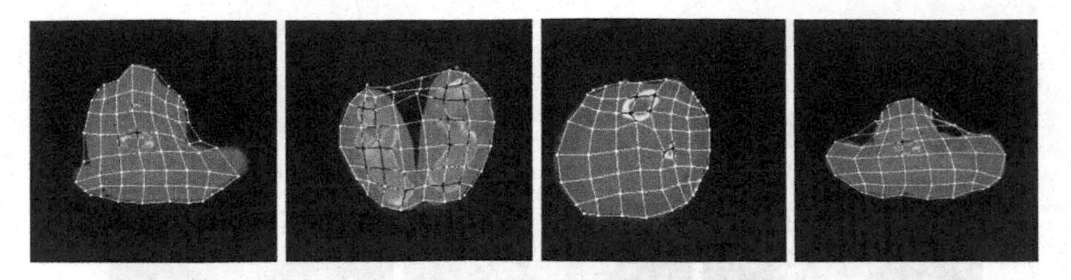

Fig. 4.6 Results obtained with the best evolved ANN and the training set of real CT images

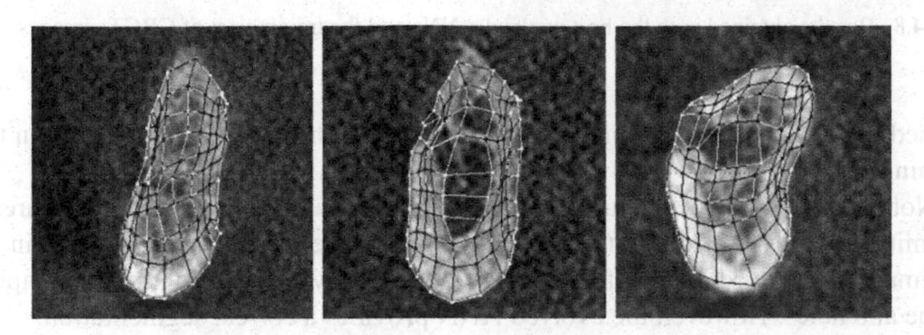

Fig. 4.7 Results obtained with the best evolved ANNs and the training set of CBCTs images

was capable to provide acceptable results, including a correct boundary detection and overcoming the presence of noise in the images. Note that the ANN also provides correct segmentations in the test images with slight different size and different positions in the whole image. In the complex parts of the segmentations, like concavities, some external nodes fall incorrectly in the background. This can be improved changing the energy parameters, increasing the TAN energy *GD* (Gradient Distance, Eq. 4.7, parameter δ), but it would deteriorate other objectives like smoothness. So, the energy parameters are always a compromise to obtain acceptable results under different shapes and conditions.

4.3.2 Segmentation of CBCT Images

The method was also tested in CBCT images, that is a more recent capture technique, variation of CT images [16]. These images have demonstrated several advantages over the conventional CT technique [17]. In particular, different images regarding the maxillofacial region to extract the osseous structures. This bone segmentation is a complex task due to the noise captured and introduced as well as the flesh region that is surrounding the different bones. In addition, the bones to be extracted present a large variable contour, making the process more complicated.

As in the previous case, we selected different images that included objects with different shapes and levels of complexity. Figure 4.7 presents the final segmentations after the training process, whereas Fig. 4.8 depicts the results obtained with the best

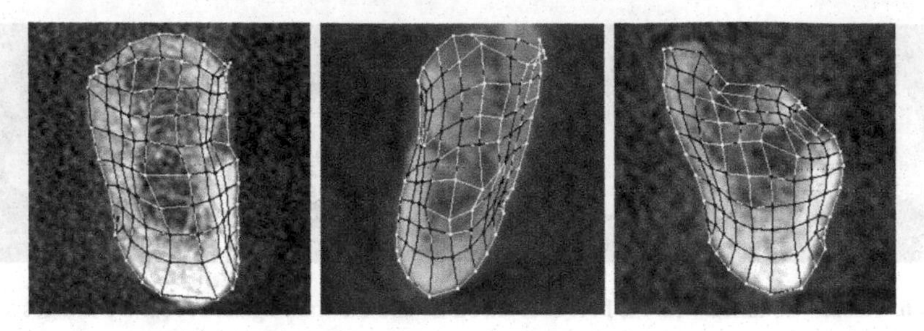

Fig. 4.8 Results obtained with the best evolved ANNs and the training set of CBCTs images

trained ANN in a set of test images with slight different shapes to those used in the training set.

Note how the ANN can situate correctly the internal nodes in the brighter areas, delimiting where the "holes" (similar gray level to the background) are, and in all the images. The third image of the test set is especially difficult as it has a complex shape and holes. However, the evolved ANN provides a correct segmentation.

4.3.3 Segmentation of Retinal Images

The method was also tested in eye fundus images. In particular, the ANN was trained to detect and segment the OD. This domain is even more complex with respect to the previous ones due to the high level of noise and other structures presented in the scene (presence of blood vessels, changes in intensities, irregular OD contour, vessels crossing the OD region . . .).

In this particular case two ad-hoc energy terms were also used, defined in [6], where a genetic algorithm was used for obtaining the final TAN with the final segmentation. These energy terms are "circularity", whose minimization favors the circular shape of the TAN (given the circular shape of the OD), and "contrast of intensities" that was designed to avoid the falling of the external nodes in the inner blood vessels. The minimization of this term tries to put the external nodes in positions with bright intensities in the inside and dark intensities in the outside. The weights or energy parameters of these two terms are $cs = 500.0$ and $ci = 500.0$, the same typical values for these terms as in [6]. In this case, we applied around 200 DE generations to obtain the optimized TAN.

Figure 4.9 shows the final segmentations obtained with the training dataset, whereas Fig. 4.10 shows the final segmentations of the test dataset and obtained with the ANN previously trained. All the images were obtained from the VARIA database of retinal images [18]. The segmentation does not have to be perfect if the final aim is the localization of the OD (as detailed in [6]), determined by the center of the final deformed TAN. So, as both figures denote, the final segmentations can

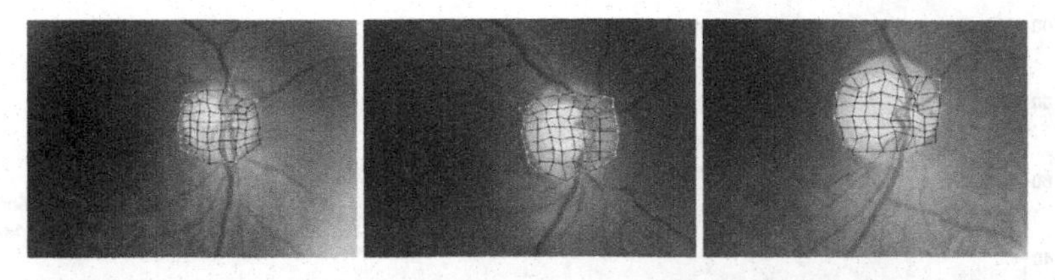

Fig. 4.9 Results obtained with the best evolved ANNs and the training set of retinal images

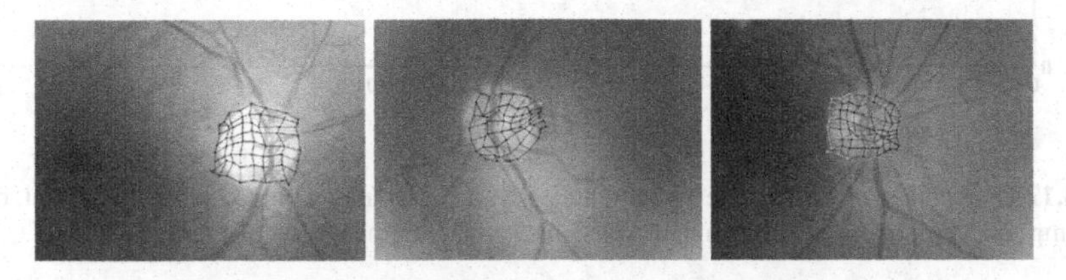

Fig. 4.10 Results obtained with the best evolved ANNs and the training set of retinal images

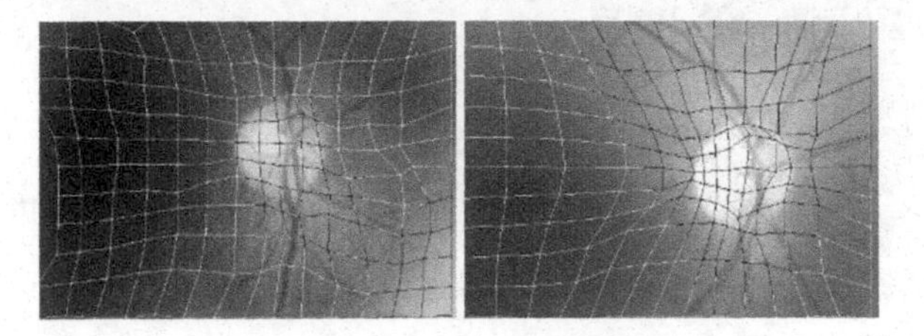

Fig. 4.11 Results obtained with the best evolved ANNs and the training set of retinal images

provide perfectly the required localization, even with different sizes and positions of the OD in the images.

To show the difference of the ANN processing with respect to a greedy search strategy, Fig. 4.11 shows the segmentation results using a greedy algorithm with two retinal images (the first images from the training and test sets). The greedy method moves iteratively all the nodes to the neighbor position with lowest energy, until no more movements can be applied. The greedy method is not able to overcome the noise in the images, as it gets stuck in few iterations in a local minimum. Nevertheless, the ANN has learned to move the nodes not necessarily in a greedy manner, optimizing the movements of the whole TAN, so the active model can progressively minimize the summed energy or fitness.

To explain why the greedy local search and the proposed method behave differently, graphics were included (Figs. 4.12 and 4.13) with the percentage of the TAN node shifts that implied a maintenance or improvement (decrement) in terms of energy, and for each step in the segmentation of the OD in the first image of Fig. 4.9. Figure 4.12 shows such percentage of movements with maintenance or improvement considering all the TAN nodes, whereas Fig. 4.13 shows the same percentage considering only the external nodes. In the graphics, the main difference between the

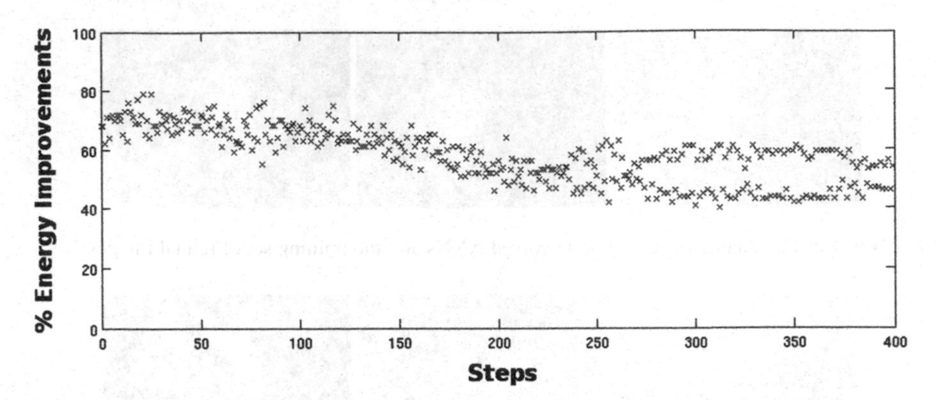

Fig. 4.12 Percentage of TANs node movements with an energy maintenance or improvement, over the temporal steps, and using the trained ANNs in the first segmentation of Figure 4.9

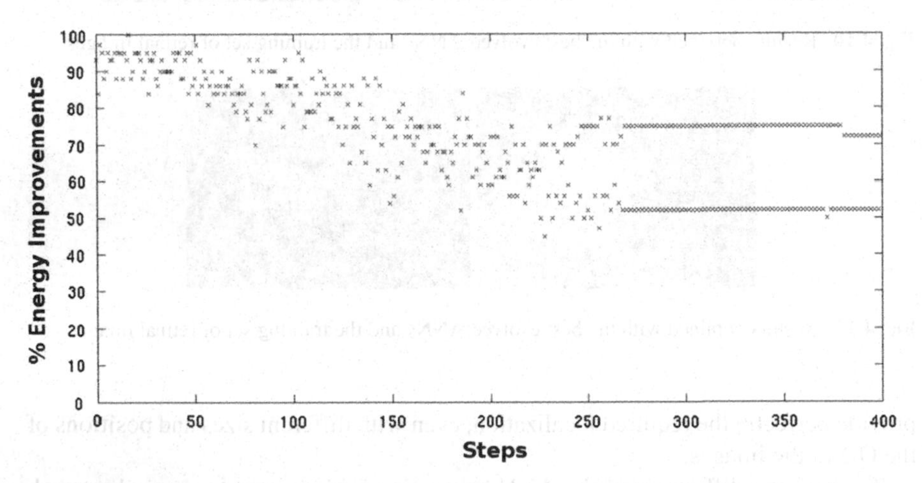

Fig. 4.13 Percentage of TANs external node movements with an energy maintenance or improvement, over the temporal steps, and using the trained ANNs in the first segmentation of Fig. 4.9

proposed method and the greedy local search is clear. Using the greedy method, all the movements of the TAN nodes imply a new position with an energy at least the same as the previous one, and better if possible (100 % in the graph). That is why, in this particular segmentation, the greedy method falls in local minima, because the nodes cannot find a better position in the neighborhood and in few steps. However, with the proposed method, the ANN learned to produce "bad" movements (an average of 50 % at the final steps), that implies worse energies in the short term, but they were suitable to find a correct segmentation in terms of the entire segmentation process. The continuous change between two different values of the percentage in the final segmentation steps (beginning around step 260, very clear with the external nodes) indicates that the ANN cannot obtain a better segmentation, since the Topological Active Net (TAN) nodes are continuously moving in the same and close areas.

4.4 Conclusions

In this work a new approach is proposed for medical image segmentation using deformable models. A deformable model, TAN, was used, integrating features of region-based and boundary-based segmentation techniques. The deformation through time was defined by an evolutionary trained ANN, since the ANN determined the movements of each of the nodes. This process was progressively repeated for all the nodes in different temporal steps until the final segmentation was obtained. Thus, the ANN provides an "emergent" segmentation, as a result of the local movements provided by the ANN and the local and surrounding energy information that the ANN receives as input. So, the trained ANN can be considered as "segmentation operators", and with a clear advantage of this approach, since that, once the ANN is trained, the emergent segmentation provided by it is faster with respect to the use of an evolutionary algorithm to directly discover the final optimized TAN, and overcoming the problems of a greedy approach in noisy images.

This approach was used in specific medical domains, that include, CT images, CBCT images and retinal images, obtaining successful results and overcoming noise problems that are presented in the images. We trained the ANN using specific training sets of images and we tested them with different test images. All the images used presented different shapes and complications and the trained ANN provided correct results with the real images.

Acknowledgments This paper has been partly funded by the Ministry of Science and Innovation through grant contracts TIN2011-25476 and TIN2011-27294 and by the Consellería de Industria, Xunta de Galicia, through grant contract 10/CSA918054PR.

References

1. Tsumiyama, K., Yamamoto, K.: Active net: active net model for region extraction. IPSJ SIG Notes **89**(96), 1–8 (1989)
2. Kass, M., Witkin, A., Terzopoulos, D.: Snakes: active contour models. Int. J. Comput. Vision **1**(2), 321–323 (1988)
3. Ansia, F., Penedo, M., Mariño, C., Mosquera, A.: A new approach to active nets. Pattern Recognit. Image Anal. **2**, 76–77 (1999)
4. Williams, D., Shah, M.: A fast algorithm for active contours and curvature estimation. CVGIP Image Underst. **55**(1), 14–26 (1992)
5. Ibáñez, O., Barreira, N., Santos, J., Penedo, M.: Genetic approaches for topological active nets optimization. Pattern Recogn. **42**, 907–917 (2009)
6. Novo, J., Penedo, M.G., Santos, J.: Localisation of the optic disc by means of GA-optimised topological active nets. Image Vis. Comput. **27**, 1572–1584 (2009)
7. Novo, J., Barreira, N., Penedo, M., Santos, J.: Topological active volume 3D segmentation model optimized with genetic approaches. Nat. Comput. **11**, 161–174 (2012)
8. Novo, J., Santos, J., Penedo, M.G.: Optimization of topological active nets with differential evolution. Lect. Notes Comput. Sci. Adapt. Nat. Comput. Algorithms **6593**, 350–360 (2011)
9. Zitzler, E., Laumanns, M., Thiele, L.: SPEA2: Improving the strength pareto evolutionary algorithm. In: EUROGEN: Evolutionary Methods for Design. Optimisation, and Control. **2002**, 95–100 (2001)

10. Novo, J., Penedo, M., Santos, J.: Evolutionary multiobjective optimization of topological active nets. Pattern Recogn. Lett. **31**, 1781–1794 (2010)
11. Novo, J., Penedo, M.G., Santos, J.: Multiobjective optimization of the 3D topological active volume segmentation model. In: ICAART: International Conference on Agents and. Artificial Intelligence. **2011**, 236–241 (2011)
12. McInerney, T., Hamarneh, G., Shenton, M., Terzopoulos, D.: Deformable organisms for automatic medical image analysis. Med. Image Anal. **6**, 251–266 (2002)
13. Price, K., Storn, R.: Differential evolution—a simple and efficient heuristic for global optimization over continuous spaces. J. Global Optim. **11**(4), 341–359 (1997)
14. Price, K., Storn, R., Lampinen, J.: Differential evolution. A Practical Approach to Global Optimization (Natural Computing Series). Springer, New York (2005)
15. Feoktistov, V.: Differential Evolution: In Search of Solutions. Springer, New York (2006)
16. Scarfe, W., Farman, A.: What is cone beam CT and how does it work? Dent. Clin. North Am. **52**, 707–730 (2008)
17. Greef, S.D., Willems, G.: Three-dimensional cranio-facial reconstruction in forensic identification: latest progress and new tendencies in the 21st century. J. Forensic Sci. **50**(1), 12–17 (2005)
18. VARIA: VARPA Retinal Images for Authentication. Varpa Web site at. http://www.varpa.es/varia.html (2008)

Chapter 5
Shape from SEM Image Using Fast Marching Method and Intensity Modification by Neural Network

Yuji Iwahori, Kazuhiro Shibata, Haruki Kawanaka, Kenji Funahashi, Robert J. Woodham and Yoshinori Adachi

Abstract This chapter proposes a new approach to recover 3-D shape from a Scanning Electron Microscope (SEM) image. When an SEM image is used to recover 3-D shape, one can apply the algorithm based on the solving the Eikonal equation with Fast Marching Method (FMM). However, when the oblique light source image is observed, the correct shape cannot be obtained by the usual one-pass FMM. The approach proposes a method to modify the original SEM image with intensity modification by introducing a Neural Network (NN). Correct 3-D shape could be obtained using FMM and NN learning without iterations. The proposed approach is demonstrated through computer simulation and validate through experiment.

Y. Iwahori (✉) · K. Shibata
Departement of Computer Science, Chubu University, 1200 Matsumoto-cho, Kasugai 487–8501, Japan
e-mail: iwahori@cs.chubu.ac.jp

K. Shibata
e-mail: shibata@cvl.cs.chubu.ac.jp

H. Kawanaka
School of Information Science and Technology, Aichi Prefectural University, 1522-3 Ibaragabasama, Nagakute-shi, Aichi 480-1198, Japan
e-mail: kawanaka@ist.aichi-pu.ac.jp

K. Funahashi
Department of Computer Science, Nagoya Institute of Technology, Gokiso-cho, Showa-ku, Nagoya 466-8555, Japan
e-mail: kenji@nitech.ac.jp

R. J. Woodham
Department of Computer Science, University of British Columbia, Vancouver, BC V6T 1Z4, Canada
e-mail: woodham@cs.ubc.ca

Y. Adachi
College of Business Administration and Information Science, Chubu University, 1200 Matsumoto-cho, Kasugai 487-8501, Japan
e-mail: adachiy@isc.chubu.ac.jp

J. W. Tweedale and L. C. Jain (eds.), *Recent Advances in Knowledge-based Paradigms and Applications*, Advances in Intelligent Systems and Computing 234,
DOI: 10.1007/978-3-319-01649-8_5,

Keywords Scanning Electron Microscope · Intensity Modification · RBF Neural Network · Fast Marching Method

5.1 Introduction

It is important to obtain the 3-D shape of the object using the SEM. Even if human can see the very small object, the shape recovery is the only way to know and evaluate the 3-D shape information.

Linear Shape from Shading proposed by Pentlabd [1] recovers shape from one shading image under the assumption of linear reflectance function. Pentland [1] uses the condition that the position of viewing point (camera) and light source is widely located to assume the linear reflectance function from the wide angle of illuminating direction.

Ikeuchi et al. [2] proposed *Shape from Occluding Boundaries* which recovers shape using the regularization with the iterative approach. The initial guess is given from the surface gradient of the points on the occluding boundaries, which becomes perpendicular to both of the contour of the boundary and the viewing direction, based on the geometrical condition. Then the iterative calculation is done using the relaxation method which is available to a simply convex closed curved surface.

While some methods have been proposed to recover the shape using the principle of the rotation of the object. *Shape from Silhouette* [3] uses multiple images through 360 degrees rotation, however, it is also unavailable to the object with local concave shape because the *Shape from Silhouette* can extract the shape of convex hull from the observed images (silhouette). The precision of the obtained result also depends on every rotating angle.

Lu et al. [4] also uses the 90 degrees rotation of an object. The method recovers shape from many images during rotation with slight angle. Since the rotation angle of the object stand of SEM is restricted, this approach is not also available. From the above situations, the previous approaches cannot be applied the SEM.

Instead of Radial Basis Function Neural Network (RBF-NN) based photometric stereo [5], Iwahori et al. [6] proposed an approach of optimization using two images observed through the rotation of the object stand. The approach is treated as the optimization problem with multiple variables based on the standard regularization theory. The optimization is applied using the Hopfield like Neural Network (HF-NN) [7]. The initial vector determined by the RBF-NN is given to the HF-NN for the optimization. Iwahori et al. [8] can have a reflectance property via many rotation images of an object and uses FMM to recover 3-D shape, however this method is not also available to SEM. Except stereo vision approaches [9–11], recent photometric approaches by Ding et al. [12] can get the surface reflectance of the object and to recover 3-D shape with FMM and this method also uses multiple images of rotation from a single view point. As another approach of monocular vision approach with a single image to recover the 3D shape, Tankus [13] introduces the iterative approach as an extension of FMM for the Lambertian object. The authors try to solve the problem of oblique light source. Although the iteration improves the recovered shapeto some extent, another approach with simple and higher accuracy is considered.

To solve the oblique light source problem, this chapter proposes a new approach to recover 3-D shape from a SEM image. When the SEM image is used to recover a 3-D shape, one can apply the algorithm based on solving the Eikonal equation with FMM. However, when the vertical and oblique light source image is observed, the correct shape cannot be obtained by the usual one-pass FMM in general. The approach proposes a method to modify the original SEM image with intensity modification via neural network so that the correct 3-D shape could be obtained with FMM. The approach has an advantage that direct shape reconstruction with intensity modification is possible without any iteration and neural network learns the mapping with actual physical environment of SEM. The proposed approach is demonstrated through computer simulation and real experiment. Section 5.2 describes the characteristics of SEM image, and Sect. 5.3 describes the basic theory to recover the shape from SEM image with oblique light source. Section 5.4 demonstrates simulation and experiment based on the proposed approach.

5.2 Characteristics of SEM Image

Here, the architecture and reflectance property of SEM and a functional form for the SEM image intensity are introduced. The architecture of the SEM is shown in Fig. 5.1.

It is noted that SEM image has the following properties [2]:

1. The brightness becomes low for the point that the surface normal is toward the viewing direction.
2. The light source and the viewing point are assumed to be located at the same position (in ideal condition) but this chapter treats the problem of the oblique light source.
3. Cast shadow does not occur.

These properties are used as the basic conditions and the chapter treats the condition (b) to recover the shape with more accurate approach.

Image intensity $I_{i,j}$ of SEM is represented in Eq. (5.1),

$$I_{x,y} = \frac{I_{min}}{\cos \theta_{x,y}} = I_{min}\sqrt{1 + p_{x,y}^2 + q_{x,y}^2} \tag{5.1}$$

where $\theta_{i,j}$ means the angle between the impinging direction of electron and surface gradient and $p_{x,y}$C$q_{x,y}$ represent the surface gradient derivative parameters of z along x and y. I_{min} represents the minimum value of the observed image $I(x, y)$, which corresponds to the surface normal to the impinging direction of electron.

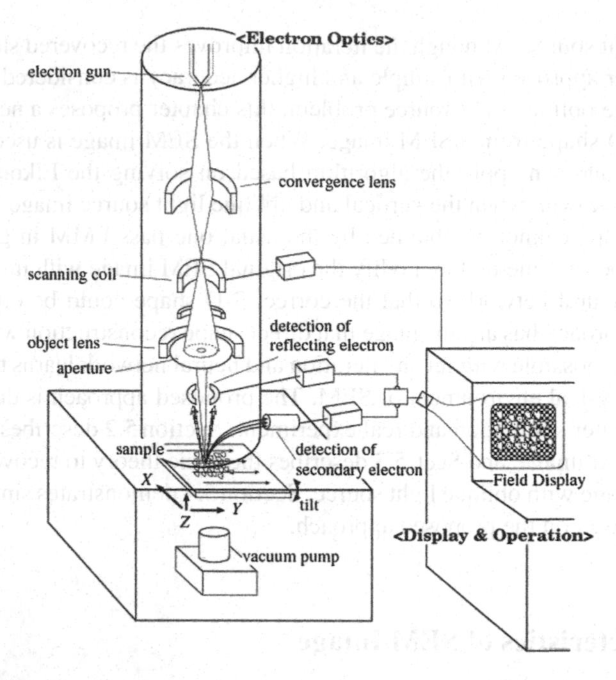

Fig. 5.1 Scanning electron microscope

5.3 Shape from SEM Image with Oblique Light Source

This section describes Fast Marching Method in Sect. 5.3.1 and modification of Eikonal equation to SEM in Sect. 5.3.2. Then how to apply for oblique light source image is shown in Sect. 5.3.3.

5.3.1 Fast Marching Method

Shape recovery using FMM [14] has been proposed by Kimmel et al. [15]. The assumptions of Kimmel et al. [15] are that the light source direction vector $\boldsymbol{s}$ is aligned with the same direction as the viewing direction vector $\boldsymbol{v}$. When $\boldsymbol{s}$ and $\boldsymbol{v}$ be (0, 0, 1), the image irradiance equation for Lambertian reflectance becomes Eq. (5.2),

$$E = \frac{1}{\sqrt{p^2 + q^2 + 1}} \tag{5.2}$$

where (p, q) are the surface gradient parameters $\partial z/\partial x$ and $\partial z/\partial y$, respectively. Eq. (5.2) is rewritten as Eq. (5.3),

$$\sqrt{p^2 + q^2} = \sqrt{\frac{1}{E^2} - 1} \tag{5.3}$$

where Eq. (5.3) is known as Eikonal equation. The fast algorithm to solve Eikonal equation is known as FMM [14] and its procedure is given as follows (See Fig. 5.2).

Step 1: Initialization—All pixels are labelled as one of three lists, $known$, $trial$, far according to the following processes (as shown in Fig. 5.2a).
1. First pixel are added to $known$ list. Z is assigned to 0.
2. Four nearest neighboring points not $known$ are labelled as $trial$ and Z is assigned to f_{ij}.
3. Other pixels are added to far list. Z is assigned as ∞.

Step 2: Select a pixel (i_{min}, j_{min}) with the minimum value of Z among $trial$ list and remove the pixel from $trial$ list and add it to $known$ list (as shown in Fig. 5.2b).

Step 3: Pixels which belong to far list among four neighboring points around (i_{min}, j_{min}) are added to $trial$ list (here, if point A is selected as a minimum point, A is added to $known$ list as shown in Fig. 5.2c).

Step 4: z of the nearest neighboring points of pixel (i_{min}, j_{min}), which belongs to $trial$ list, is calculated and registered (as shown in Fig. 5.2d).

Step 5: When the pixels which belongs to $trial$ exists, return Step 2 else exit.

Eikonal equation is derived and solving this equation by FMM can recover the shape of Lambertian object from one light source. However, the surface reflectance is limited to Lambertian reflectance based on the form of Eikonal equation. It means that the approach of Kimmel et al. [15] itself cannot be applied to specular reflectance.

5.3.2 Modification to SEM

Transforming Eq. (5.1) results in Eq. (5.4),

$$\sqrt{p_{i,j}^2 + q_{i,j}^2} = \sqrt{\frac{I_{i,j}^2}{I_{min}^2} - 1} \tag{5.4}$$

where the right term of this equation can be replaced to the function of SEM image. This means that FMM is applied to SEM image for this Eikonal equation. The initial depth is given at the initial point which has the local minimum point for SEM image.

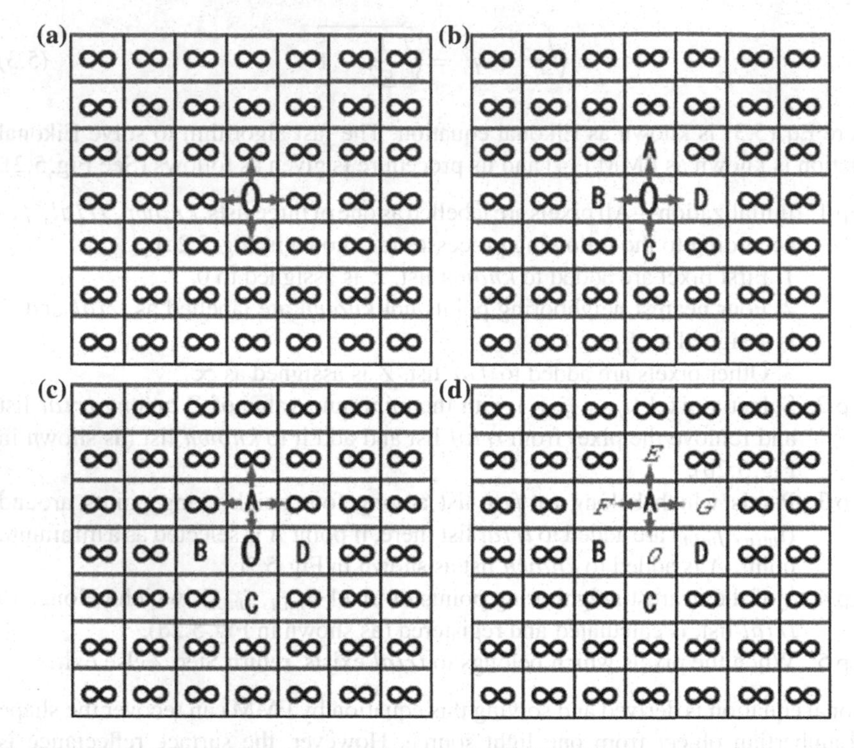

Fig. 5.2 FMM. (**a**) Initialization *known*{0}, *far*{∞}, *trial*{ }. (**b**) Determining 4 nearest neighbour points *known*{0}, *far*{∞}, *trial*{A,B,C,D}. (**c**) Selecting minimum points among ABCD *known*{0,A}, *far*{∞}, *trial*{B,C,D}. (**d**) Determining temporal value of 4 nearest points known{0,A}, *far*{∞}, *trial*{B,C,D,E,F,G}

5.3.3 Intensity Modification for Oblique Light Source Image

Scanning Electron Microscope (SEM) image with oblique light source is observed based on the effect of location of electron detector. In this case, the exact shape cannot be obtained with FMM because FMM assumes the circular symmetric reflectance function obtained from the frontal illuminating direction. The proposed approach modifies the intensity of image obtained from the oblique light source direction to the frontal light source direction by using FMM under the better condition. The intensity modification is performed by nonlinear transformation and to treat the physical and synthesized input/output using a sphere object, neural network is used to modify the intensity distribution of a target object after learning the input/output relation using a sphere object. Although Neural Network (NN) is the only choice to perform this purpose, the reason is that a NN can be applied to perform the nonlinear functional approximation with high ability.

Here, a RBF-NN proposed by Chen et al. [16] is one choice suitable for many applications. In particular, it has been widely used for strict interpolation in

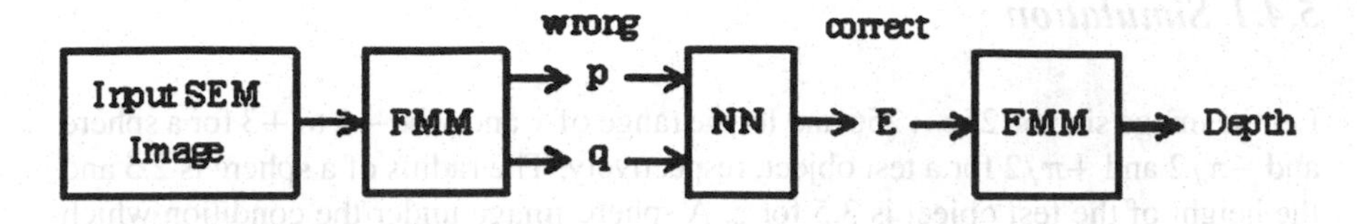

Fig. 5.3 Intensity modification of proposed approach

multidimensional spaces. It is argued that RBF-NN often can be designed in a fraction of the time it takes to train standard feed-forward networks. RBF-NN is claimed to work well when many training vectors are available and RBF-NN represents nonlinearity via the chice of basis function. Gaussian is not the only choice of RBF but it is the choice widely used and the one used here.

Here, Intensity modification with FMM and this RBF-NN is proposed as shown in Fig. 5.3.

The procedures are as follows:

1. FMM is applied to the original image of a sphere object just like oblique light source. The image has the local minimum point and this point is used as the initial point and z-distribution is recovered with FMM.
2. The surface gradient parameters (p, q) are calculated with the numerical difference of z for a sphere object. However this calculated (p, q) value is not correct and it is used to the input of NN while the correct image intensity at the corresponding point of the wrong (p, q) is used for the output of NN for the learning. The intensity modification NN is learned for the ideal (synthesized) SEM image of a sphere with the mapping of the wrong (p, q) for the original sphere to the intensity E of a synthesized sphere at the corresponding point. Intensity for the test object can be modified using the mapping of learned NN.
3. FMM is applied again to recover the reliable depth map.

While FMM is applied for the test object in the same way as a sphere and its modified intensity image is obtained via the learned NN for a sphere. Here, numerical difference of z is applied for the test object to get the wrong (p, q) as the input to NN. After the modification of image intensity, z-distribution with the higher accuracy is obtained via only applying FMM again. The proposed approach does not need some iterations for applying FMM proposed in [13].

5.4 Experiments

Computer simulation and real experiment using SEM were done for the evaluation of the proposed approach. Computer simulation is described in Sect. 5.4.1 and the experiments for real SEM image is described in Sect. 5.4.2.

5.4.1 Simulation

Let the image size be 256×256 and let the range of x and y be -3 to $+3$ for a sphere and $-\pi/2$ and $+\pi/2$ for a test object, respectively. The radius of a sphere is 2.5 and the height of the test object is 3.5 for z. A sphere image under the condition which corresponds to the oblique light sourceis shown in Fig. 5.4a. FMM is applied to the image from the initial point and z distribution is obtained. NN is constructed with the mapping of the wrong (p, q) calculated from the z distribution obtained by the first FMM to the correct intensity E at the corresponding point under the condition of the frontal illumination. Target image to learn with NN is shown in Fig. 5.4b as an example of Lambertian case.

FMM is applied to the image of test object in Fig. 5.5a and (p, q) is calculated from z and input to the modified NN. The output of NN is E and FMM is applied again to the modified image shown in Fig. 5.5b.

The learning state of NN to modify the image intensity with different learning epochs are shown in Fig. 5.6a–c. Figure 5.6 shows the learning state from 100 to 300, and 300 learning epochs are sufficient to learn the modification NN.

The modified image is generated by using this learned NN. Spread parameter for RBF-NN was 0.5 for the distribution of (p, q) and the error goal was set to be 10^{-6} with the learning state in Fig. 5.6.

Figure 5.7a is the recovered result for a sphere model for the Fig. 5.4a using FMM and the modified NN. FMM is first applied to Fig. 5.4, (p, q) is calculated from z

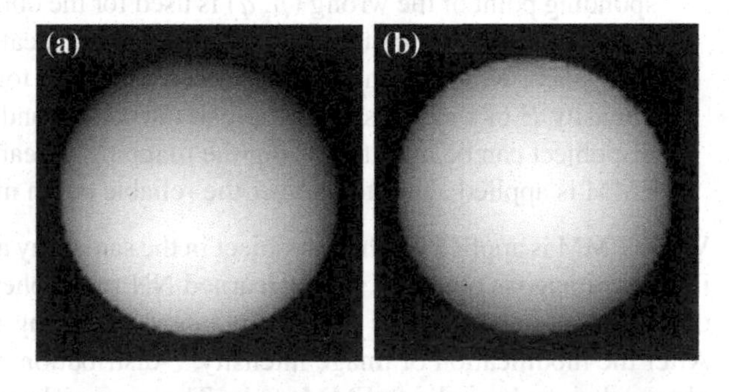

Fig. 5.4 Lambertian sphere. (**a**) Input image. (**b**) Target image

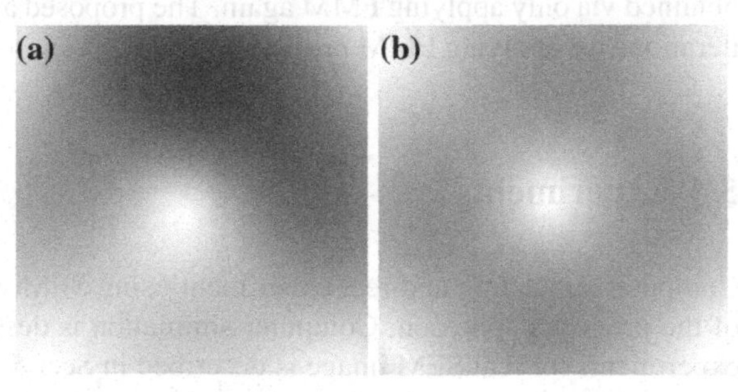

Fig. 5.5 Test object image. (**a**) Before modification. (**b**) After modification

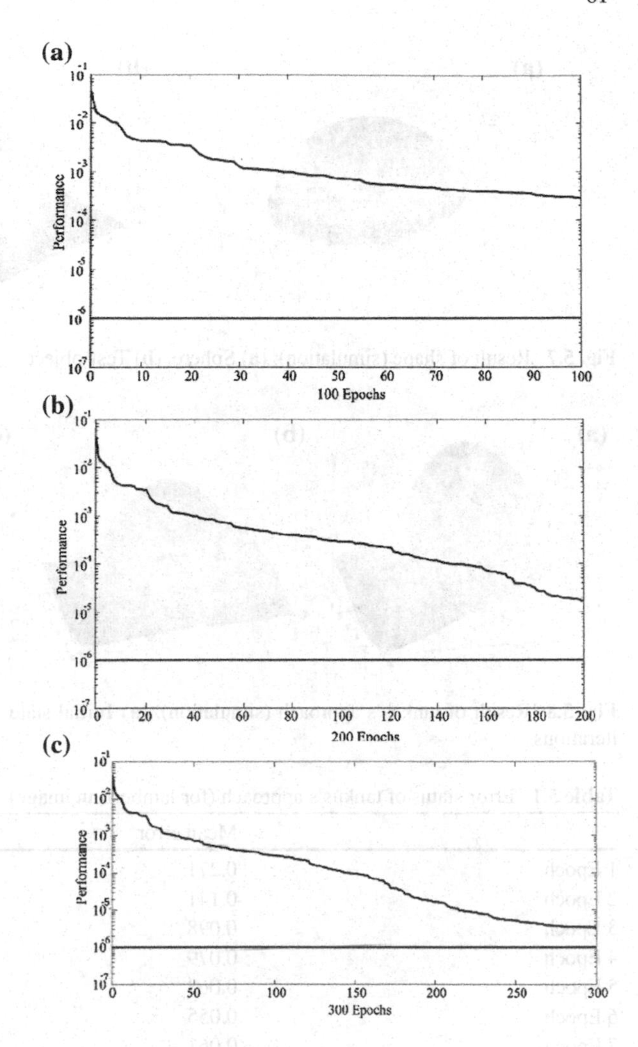

Fig. 5.6 Learning status of neural network. (**a**) 100 Epochs. (**b**) 200 Epochs. (**c**) 300 Epochs

obtained from FMM. Then modified E and FMM is applied again to the modified image shown in Fig. 5.4b.

Another recovered result for test object is shown in Fig. 5.7b. This result was also obtained using the modified NN which was learned with a sphere model.

For the comparison, the results by Tankus's approach [13] are shown in Fig. 5.8. This approach decreases the error via iteration and the status of error is shown in Table 5.1. Here it is assumed that the range value of E is between 0 and 1.

Several numbers of iterations are necessary to satisfy the condition that the shape is a little bit changed but almost converged, but results have still some inclined shape in comparison with the proposed approach. The evaluation of error for the comparison with approach Tankus et al. [13] is shown in Table 5.2.

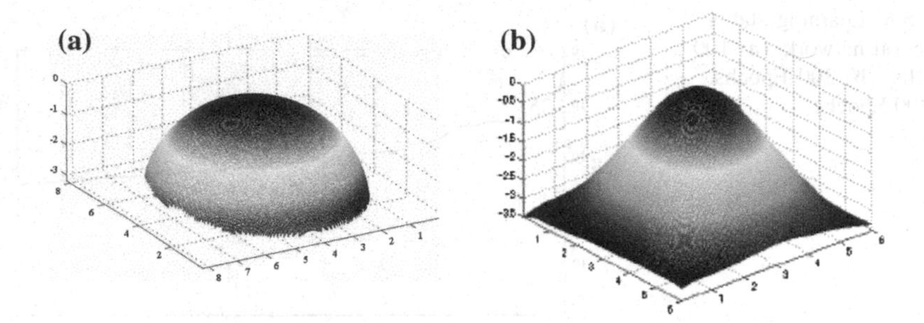

Fig. 5.7 Result of shape (simulation). (**a**) Sphere. (**b**) Test object

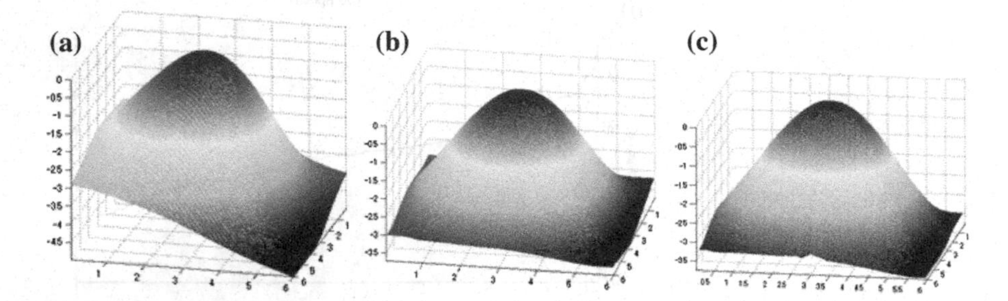

Fig. 5.8 Result of tankus's approach (simulation). (**a**) Initial state. (**b**) Four iterations. (**c**) Eight iterations

Table 5.1 Error status of tankus's approach (for lambertian image)

	Mean error	Max error
1 Epoch	0.271	0.581
2 Epoch	0.141	417
3 Epoch	0.098	0.354
4 Epoch	0.079	0.235
5 Epoch	0.063	0.219
6 Epoch	0.055	0.201
7 Epoch	0.061	0.289
8 Epoch	0.078	0.342

Error evaluation was done for this height. The mean error and maximum error for a recovered sphere were 0.008 and 0.031 respectively, while the mean error and maximum error for the test object was 0.011 and 0.036, respectively. Error for test object was a little bit larger than that of sphere. The error was smaller for both objects in comparison with Tankus et al. [13] and it is considered that there are some points which were not learned with NN because of the lacked information of self-shadow problem for the sphere generated by the simulation. It was confirmed that intensity modification was correctly done and the corresponding z distribution with higher accuracy was obtained via NN and FMM for the test object.

Table 5.2 Comparison of test object error verses tankus

	Mean error	Max error
Proposed approach	0.011	0.036
Approach	0.055	0.201

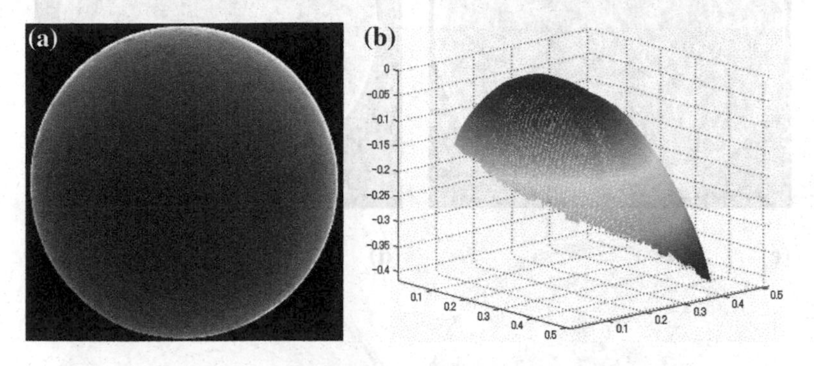

Fig. 5.9 Result without modification. (**a**) Original image. (**b**) Recovered result

5.4.2 Real SEM Image

A sphere with stainless steel with radius 0.5mm is used and the SEM image is shown in Fig. 5.9a.

FMM is applied without any further processing is shown in Fig. 5.9b. The real sphere SEM image (which corresponds to the oblique light source) and the ideal sphere image generated with the reflectance function (which corresponds to the frontal illumination) are used for the NN learning. The real sphere in Fig. 5.9a is recovered with FMM and the mapping of (p, q) calculated from its z to the true intensity E at the corresponding point is used for the intensity modification. FMM is applied to the modified image again to recover the shape for the evaluation.

The image used for the learning of NN is shown in Fig. 5.10b. The corresponding contour images for Fig. 5.10a, b are shown in Fig. 5.10c, d, respectively. The value of (p, q) was calculated from the z which was obtained for Fig. 5.10a with FMM. This value of (p, q) is input and E in Fig. 5.10b at the point with its value of (p, q) is output for NN learning.

The result after the modification with NN and the result shape with FMM for the modified image are shown in Fig. 5.11.

It is confirmed that the result from the original image (without modification) has larger error for the original radius of 0.25 mm, while the result from the modification image shown in Fig. 5.11b has improved the accuracy with the mean error of 0.002 and the maximum error of 0.035. The real SEM image of a sphere object does not lack the part in self-shadow region like simulation. This makes it possible to reduce the error to learn the mapping of NN at every point in the real experiments for the case in simulation.

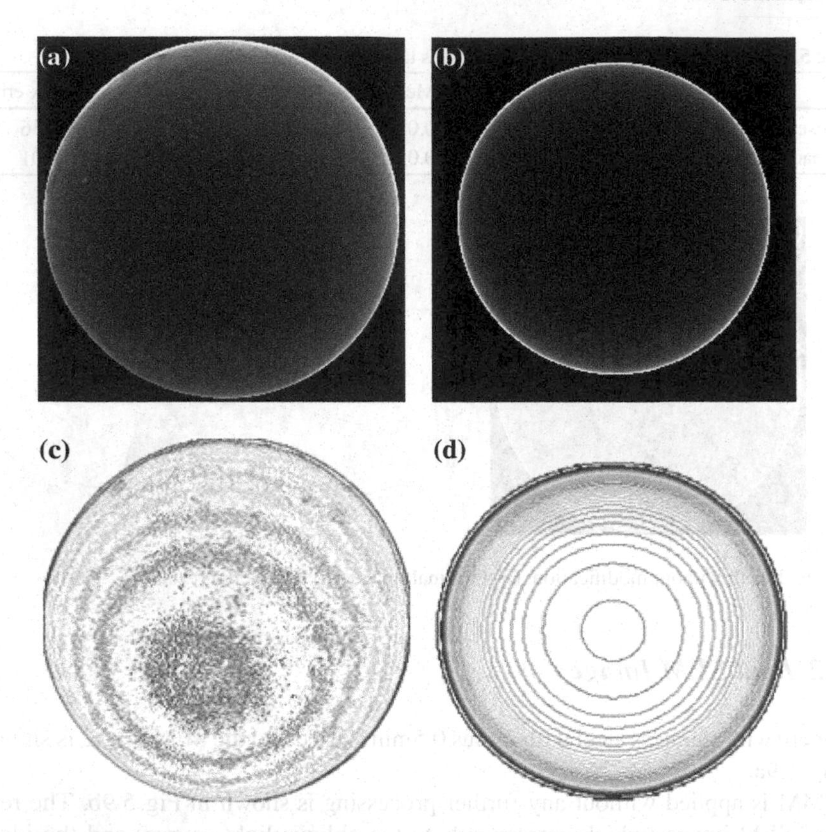

Fig. 5.10 Sphere image for NN learning. (a) Input image (before modification). (b) Image used for NN learning. (c) Contour image (before modification). (d) Contour image (for NN learning)

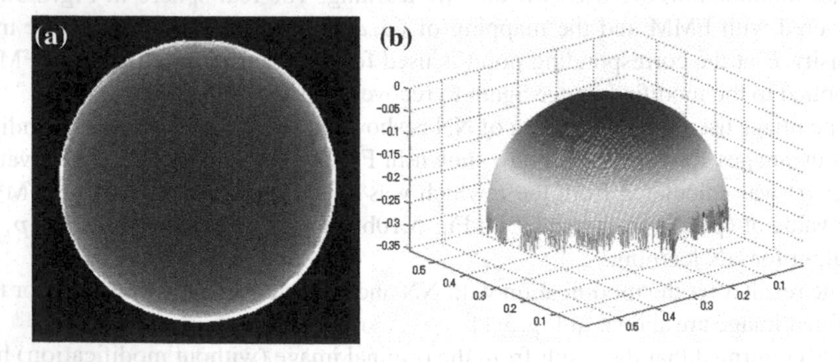

Fig. 5.11 Result after modification. (a) Modification image. (b) Recovered result

The experiment for another test object was also done. The input image for the object of solder is shown in Fig. 5.12a and recovered result without Modification is shown in Fig. 5.12b. While the modification image is shown in Fig. 5.13a and its

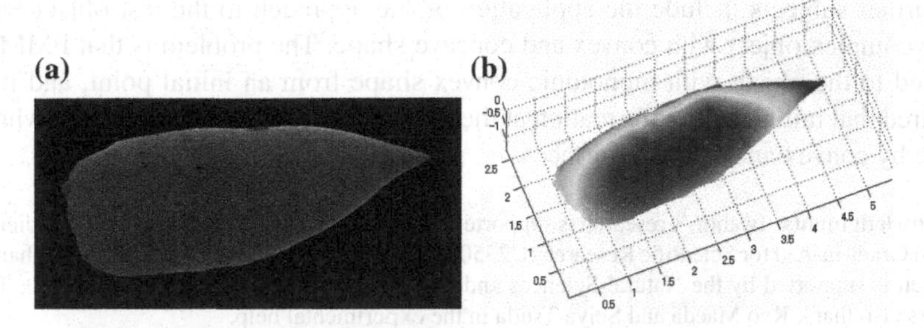

Fig. 5.12 Result without modification. **a** Original image. **b** Recovered result

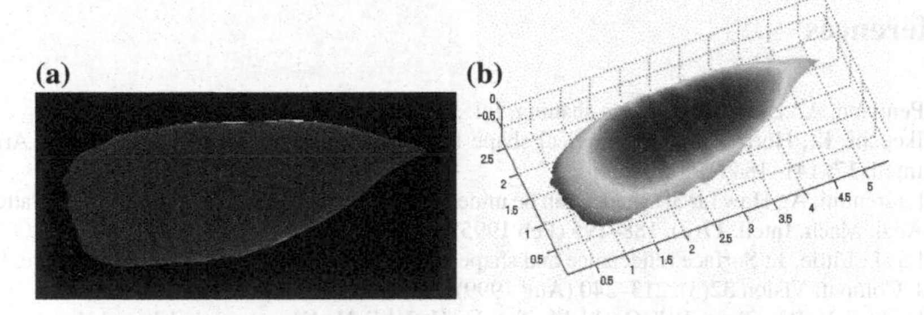

Fig. 5.13 Result after modification. **a** Modification image. **b** Recovered result

recovered result is shown in Fig. 5.13b. The convex part of the top slips out of place in Fig. 5.13a while the corresponding part fits one perfectly in Fig. 5.13b. Although the quantitative analysis is difficult because the true shape is unknown but the recovered result of Fig. 5.13b gives qualitatively better result than that of Fig. 5.12b. This is the concluded remark of the chapter.

5.5 Conclusion

This chapter proposed a new approach to modify the image intensity with NN for a observed SEM image like the image from the oblique light source. A sphere is used for a calibration and learning with NN and another test object can be recovered using FMM for a modification image. The modification is done with neural network for a sphere object. It is shown that the proposed approach can recover the correct shape with higher accuracy. Although SEM images for a sphere is used for a real sphere image for taking input data and the synthesized sphere image for taking output of the NN. This makes it possible to improve the accuracy for the recovered shape.

Further subjects include the application of the approach to the test object with more complex object with convex and concave shape. The problem is that FMM is applied to the object with monotonic convex shape from an initial point, and it is required that initial depth information at the multiple initial points recover the whole shape by convex and concave surfaces.

Acknowledgments Iwahori's research is supported by Japan Society for the Promotion of Science (JSPS) Grant-in-Aid for Scientific Research (C23500228) and Chubu University Grant. Woodham's research is supported by the Natural Sciences and Engineering Research Council (NSERC). The authors also thank Ryo Maeda and Seiya Tsuda in the experimental help.

References

1. Pentland, A.: Linear shape from shading. Int. J. Comput. Vision **4**, 153–162 (1990)
2. Ikeuchi, K., Horn, B.K.P.: Numerical shape from shading and occluding boundaries. Artif. Intell. **17**, 141–184 (1981)
3. Laurentini, A.: How far 3D shapes can be understood from 2D silhouettes. IEEE Trans. Pattern Anal. Mach. Intell. **17**(2), 188–195 (Feb 1995)
4. Lu, J., Little, J.: Surface reflectance and shape from images using a collinear light source. Int. J. Comput. Vision **32**(3), 213–240 (Aug 1999)
5. Iwahori, Y., Woodham, R.J., Ozaki, M., Tanaka, H., Ishii, N.: Neural network based photometric stereo with a nearby rotational moving light source. IEICE Trans. Info. Syst. **E80-D**(9), 948–957 (1997)
6. Iwahori, Y., Kawanaka, H., Fukui, S., Funahashi, K.: Obtaining shape from scanning electron microscope using hopfield neural network. J. Intell. Manuf. **16**(6), 715–725 (2005)
7. Hopfield, J.J., Tank, D.W.: "Neural" computation of decisions in optimization problems. Biol. Cybern. **52**, 141–152 (1985)
8. Iwahori, Y., Nakagawa, T., Woodham, R.J., Kawanaka, H., Fukui, S.: Shape from self-calibration and fast marching method. In: 19th International Conference on Pattern Recognition (ICPR 2008), pp. 1–4 (2008)
9. Ojima, H., Murakami, Y., Otsuka, H., Sasamoto, Y., Zhou, L., Shimizu, J., Eda, H.: 3D data acquisition and reconstruction from SEM stereo pairs (in Japanese), J. Jpn. Soc. Precis Eng. **75**(6), 773–777 (2009)
10. Jiang, J., Sakai, S.: 3-Dimensional measurement of fracture surface using SEM via the combination of stereo matching and integrating secondary electron intensity (in Japanese). J. Jpn. Soc. Mech. Eng. **68**(666), 300–306 (2002)
11. Kamata, K.: 3D Shape Recovery From SEM Image Using Shading Information (in Japanese), 21st Sens. Forum. C, pp. 279–284 (2004)
12. Ding, Y., Iwahori, Y., Nakamura, T., Lifeng He, Woodham, R.J., Itoh, H.: Neural network implementation of image rendering via self-calibration. J. Adv. Comput. Intell. Intell. Inform. **14**(4) 344–352 (2010)
13. Tankus, A., Sochen. N., Yeshurun. Y.: Shape-from-Shading By Iterative Fast Marching For Vertical and Oblique Light Sources. Geom. Prop. Incomplete Data, pp. 237–258 (2005)
14. Sethian, J.A.: A fast marching level set method for monotonically advancing fronts. Proc. Nat. Acad. Sci. **93**(1), 5–28 (1998)
15. Kimmel, R., Sethian, J.A.: Optimal algorithm for shape from shading and path planning. JMIV **14**, 237–244 (2001)
16. Chen, S., Cowan, C.F.N., Grant, P.M.: Orthogonal least squares learning algorithm for radial basis function networks. IEEE Trans. on Neural Networks **2**(2), 302–309 (1991)

Chapter 6
Fuzzy Evidence Reasoning and Navigational Position Fixing

Włodzimierz Filipowicz

Abstract In the traditional approach to position fixing, the navigator exploits available uncertain measurements and various system indications. Thanks to his nautical knowledge, the position is fixed and evaluated. The final results are rather intuitive than proven by limited data processing. Mathematical apparatus based on probability theory and series of assumptions is not flexible enough to include knowledge and ignorance into a position fixing calculation scheme. Limited ability is available regarding the fix accuracy evaluation. In the chapter Mathematical Theory of Evidence (MTE) is exploited in order to extend foundations for implementation of new approaches. The theory, extended for the fuzzy environment, creates new opportunities enabling modelling and the solving of problems with uncertainty. The concept of using a new basis in nautical science, for position fixing in particular, was presented by the author in his previous publications. Herein, a comprehensive introduction to the platform is presented, evidence assignment transformation is depicted within the context of solving the position fixing problem.

Keywords Evidence Representation · Belief Structures · Normalization · Navigation

6.1 Introduction

Authors of technical report devoted to the theory of evidence [1] pointed at Bayesian networks and fuzzy sets as major tools applied for reasoning under uncertainty. They also put emphasis on scarce application of Dempster-Shafer to solve any concrete problem of this kind. The presented chapter shows new and real field of applicability of a theory of evidence which can be extended for use within a possibilistic platform.

W. Filipowicz (✉)
Faculty of Navigation, Gdynia Maritime University, Gdynia, Poland
e-mail: w.filipowicz@wn.am.gdynia.pl

J. W. Tweedale and L. C. Jain (eds.), *Recent Advances in Knowledge-based Paradigms and Applications*, Advances in Intelligent Systems and Computing 234,
DOI: 10.1007/978-3-319-01649-8_6,

Practical navigation exploits graphical and analytical methods; its scientific background is based upon probability theory. The basis is enough to define distributions of random variables that are assumed to be of measured value. It also enables a priori evaluation of fixes taken according to certain schemata since accuracy is calculated with formulas designated for selected schedules of observations taking into account the constellation of landmarks and approximate measurements error.

The traditional way of position fixing takes advantage of available measurements, their approximate random distributions and diversification of observations once the analytical approach is used. The main disadvantage of the approach is the lack of a built-in universal method of the fix a posteriori evaluation.

Expectations regarding flexibility of the upgraded models are greater. Items that should affect fixed position should also include the kind of distributions of measurements taken with a particular navigational aid and discrepancies in the parameters of such distributions. It is popular to state that the mean error of a bearing taken with radar is interval valued within specified range usually written as [$\pm 1°$, $\pm 2.5°$] [2]. The presented evaluation of the mean error appears as a fuzzy figure and as such, fuzziness should be accepted and taken into account during computations. Subjective assessment, also in form of linguistic terms, of each observation should be accepted and processed. Empirical distributions are also supposed to be included in calculations. The most important thing is the embedded ability for objective evaluation of the obtained fix along with measures indicating the probability of its locations within the surrounding area. Meeting the above stated expectations is impossible with traditional formal apparatus. Its ability is almost exhausted in the considered applications. The author's recent research and published works devoted to a new platforms and modern environments put attention on Mathematical Theory of Evidence (MTE) that delivers a wide range of new opportunities. The main advantage of the proposed method is ability of thorough a posteriori analysis of the fixed position that is supposed to be carried out in geodesy and navigation. In order to validate and illustrate the approach software intended for position fixing with Mathematical Theory of Evidence (FIXMTE) has been developed. An example of its output is presented in the chapter.

Following sections contain description of the uncertain nautical evidence and its representation. Belief assignments are presented for example pieces of evidence related to two observations. Further short description of the graphic method of making the fix is enclosed. Next, the chapter contains discussion on belief structures normalization and combination. Main feature of widely used concepts are depicted and the author's way of transformation introduced.

6.2 Uncertain Evidence and its Representation

In the possibilistic approach uncertain evidence is represented with fuzzy sets and masses of confidence attributed to these sets. An hypothesis frame is also considered. Relations between the hypothesis (Ω_H) and evidence spaces (Ω_E) are encoded into evidence representation. Relations can be of binary or fuzzy types. Fuzzy sets embrace grades expressing possibilities of belonging of consecutive hypothesis items to the sets related to each piece of evidence [3].

As already mentioned each of the fuzzy sets has an assigned credibility mass also considered as a degree of belief. Therefore, fuzzy evidence mapping consists of "fuzzy set—mass of confidence" pairs. The mapping is described by Eq. 6.1 [4].

$$\mathrm{m}(e_i) = \{(\mu_{i1}(x_k),\ f(e_i \to \mu_{i1}(x_k))),\ \ldots,\ (\mu_{in}(x_k),\ f(e_i \to \mu_{in}(x_k)))\} \quad (6.1)$$

In the presented application evidence representation consists of pairs: fuzzy vectors representing locations of a set of points within sets related to each piece of evidence—degrees of credibility assigned to these vectors. The degree of confidence reflects the probability of identifying a true isoline or indication of being located within the given strip area, or inside belts intersection regions. Appropriate imprecise values are obtained based on the statistical distribution of observations gathered during the investigations.

Fuzzy sets are represented by membership functions that reflect the relationship between two universes or evidence and hypothesis spaces. It is assumed that each piece of evidence is accompanied by a set of areas, ranges, therefore membership functions reflect the relation between elements belonging to hypothesis space and sets related to elements of the evidence frame. Membership function converts the hypothesis space into power set of [0, 1] interval.

In position fixing, fuzzy sets are interpreted in the way shown in Figs. 6.1 and 6.2. In Fig. 6.1, the intersections of two imprecise lines of positions are presented. It is

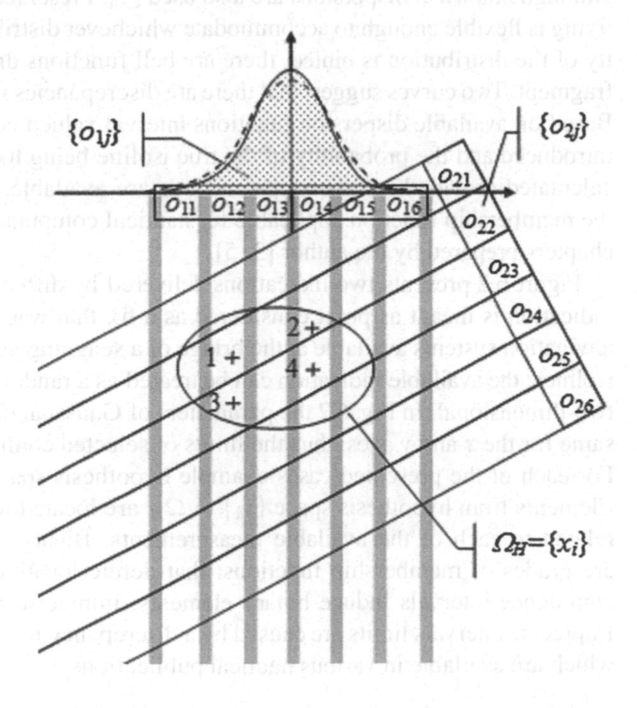

Fig. 6.1 Evidence related sets and intersection of *two line* of position along with hypothesis space elements

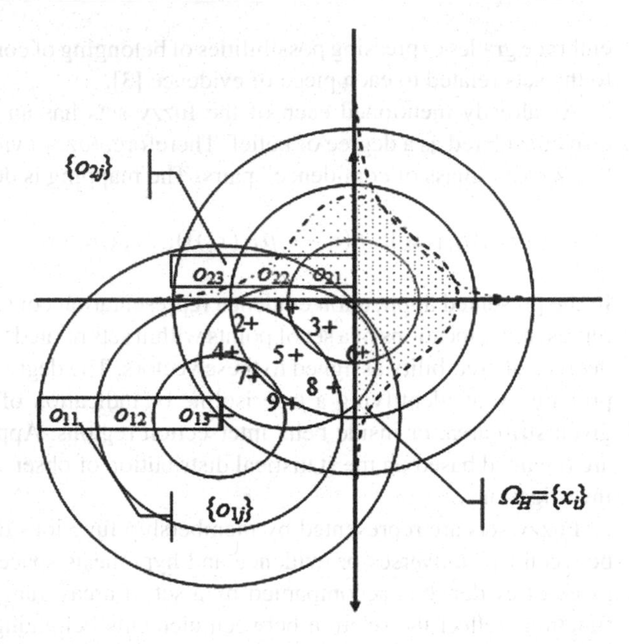

Fig. 6.2 Indicated positions related sets along with hypothesis space elements

widely assumed that isolines are randomly distributed around this in direct reference to the measurement. In most cases the distribution is considered as a normal one, although empirical dispersions are also used [2]. Presented approach to the position fixing is flexible enough to accommodate whichever distribution. In Fig. 6.1 normality of the distribution is hinted, there are bell functions drawn on top of the isoline fragment. Two curves suggest that there are discrepancies in distribution estimations. Based on available dispersion functions interval valued confidence intervals can be introduced and the probability of the true isoline being located within them readily calculated provided membership functions are available. For discussion regarding the membership function applicable for nautical computation schemes see previous chapters prepared by the author [3, 5].

Figure 6.2 presents two indications delivered by different navigational aids. The indication is meant as point considered as a fix that was obtained from one of the navigation systems available at the bridge of a seagoing vessel. Very much like with isolines, the available indication can be treated as a random variable, this time being two-dimensional. In Fig. 6.2 the parameters of Gaussian distributions are almost the same for the x and y axes, thus the limits of selected confidence ranges are circular. For each of the presented cases example hypothesis spaces (Ω_H) are also shown. Elements from hypothesis space $\{x_k\} = \Omega_H$ are located within reference sets $\{o_{ij}\}$ related to each of the available measurements. Binary or fuzzy valued locations are grades of membership functions: that define location vectors. Crisp limits of confidence intervals induce binary elements, imprecise limits stipulate fuzziness. Imprecise intervals limits are caused by a discrepancy in distribution characteristics, which are available in various nautical publications.

Hereafter, based on the above rationale, function $f(e_i \rightarrow \mu_{ij}(x_k))$ that defines degree of belief (also referred to as a mass) assigned to each location vector that is directly related to jth confidence interval, is assumed known for the hypothesis set with respect to ith measurement or indication [6, 7].

The set of expressions listed in Eqs. 6.2–6.5 present limitations that are usually applied to evidence representation. At first the specified conditions exclude empty sets (Eq. 6.2). Second condition expressed by Eq. 6.3 in constraints specification stipulates a normality of fuzzy sets. Normal sets should include highest grade(s) equal to one. Apart from these two limitations, typical for fuzzy mapping, additional requirements stipulated by Eqs. 6.4 and 6.5, regarding greater than zero masses and their total value, are to be observed.

Evidence mapping (specified by Eq. 6.1), also called the basic probability assignment [8], can be considered as belief structure, provided conditions expressed by Eqs. 6.2–6.5 are satisfied [9]. It is widely assumed that the belief structure is to be explored in order to draw justifiable conclusions [10, 11]. Belief structures processing with null generated operator may result in assignment for which specified conditions are not observed. These kind of assignments are to be normalized. In the reference [11] the author suggests that the main reason for the normalizing process is avoiding belief being greater than plausibility measure. Both values are treated respectively as lower and upper limits of interval valued support probability [12]. Further on it will be proven that for evidence representations involving normal fuzzy sets (see Eq. 6.4 in the specified conditions set) belief does not exceed plausibility when position fixing problem is considered.

$$\mu_{ij}(x_k) = g(\{x_k\} \rightarrow o_{ij} \in \Omega_E) \neq \emptyset \tag{6.2}$$

$$m_{ij} = f(e_i \rightarrow \mu_{ij}(x_k)) \geq 0 \tag{6.3}$$

$$\max_k \mu_{ij}(x_k) = 1 \tag{6.4}$$

$$\sum_{j=1}^{n} f(e_i \rightarrow \mu_{ij}(x_k)) = 1 \tag{6.5}$$

An example of evidence mapping is presented in Table 6.1. Presented assignment refers to the scheme shown in Fig. 6.1. In this case the hypothesis space embraces four points, that is: $\Omega_H = \{x_1, x_2, x_3, x_4\}$. Each considered point can be potentially treated as fixed position. The truth of the proposition is to be proved based on plausibility and belief values that are measures exploited in MTE [13, 14].

In the example, due to the particular allocation of the hypothesis frame points, sets related to each piece of evidence can be reduced to the following items: $e_1 \rightarrow \{o_{12}, o_{14}\}$ and $e_2 \rightarrow \{o_{22}, o_{23}\}$.

Function enabling calculation of membership grades, takes the form of Eq. 6.6. It can be read that location grades are degrees of inclusion of hypothesis points within the evidence frame. In the formula $C = 1$ for binary approach, $C \in [0, 1]$ when fuzzy membership is engaged.

Table 6.1 Evidence mapping example

l.v.[a]	x_1	x_2	x_3	x_4	m
$\mu_{o12}(x_k)$	{ 1	0	1	0 }	$m_{o12} = 0.2$
$\mu_{o14}(x_k)$	{ 0	1	0	1 }	$m_{o14} = 0.8$
$\mu_{o22}(x_k)$	{ 1	0.8	0	0 }	$m_{o22} = 0.3$
$\mu_{o23}(x_k)$	{ 0	0.2	1	1 }	$m_{o23} = 0.7$

[a] l.v.—stands for location vector

$$\mu_{ij}(x_k) = \begin{cases} C & \text{if } x_k \in o_{ij} \\ 0 & \text{otherwise} \end{cases} \tag{6.6}$$

Data presented in Table 6.1 indicates that points x_1 and x_2 are situated within area o_{12}, while points x_3 and x_4 are located inside area o_{14}. These are categorized as a binary type, particular points are situated within or outside a given range. This type of membership is justified for any location close to the middle section of the considered range [6]. Membership of the second point inside area o_{22} is to be treated in quite a different, fuzzy way [15]. It is situated close to the border of the area. Subsequently its location should be partially within adjacent ranges. To some extent it is located within range o_{22} and partly inside the adjacent one. From Table 6.1 it can be seen that this point membership grade for area o_{22} is 0.8, and for range o_{23} is equal to 0.2.

It should be also noticed that the data gathered in Table 6.1 composes of two correct belief structures since all conditions stipulated by Eqs. 6.2–6.5 are observed for the first and second pairs of rows. Each structure is related to a separate piece of evidence, which in turn is defined by a measurement and knowledge regarding its accuracy. Subjective evaluation of each observation is usually a part of a belief structure. The evaluation expresses credibility attributed to particular observation. Measurements are taken to pronounced landmarks but it is quite often that vague unclear objects are observed. Ability of diversification of available observations is very practical and is recommended to be embedded into position fixing scheme [3, 6]. Nonetheless herein it was deliberately dropped in order to emphasize Bayesian, probabilistic evidence representation versus possibilitic approach.

6.2.1 Position Fixing and Evidence Combining

Single belief structure containing encoded data regarding isoline referring to a single measurement is not sufficient to make a fix. In order to achieve the goal one has to engage at least two lines of position. Figure 6.3 presents graphical position fixing scheme with three imprecise distances. Inaccuracy is hinted with bell functions drawn at each isoline. The point that should be considered as a fix is marked with a cross. Small circle around it reflects estimated accuracy of the fix. It should be noted that nautical intuition is exploited when making a fixed position especially when erratic observations are given, which is very often the case [2].

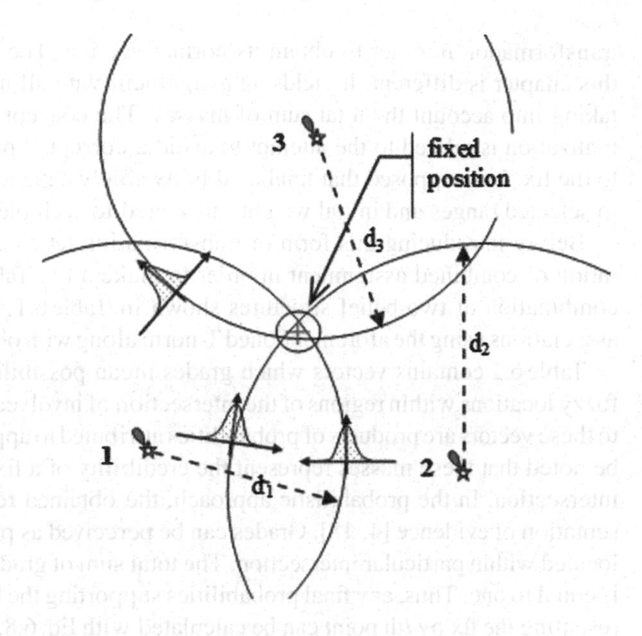

Fig. 6.3 Scheme of position fixing with three imprecise distances

In MTE available evidence is encoded into belief structures and their combination is carried out [5, 9]. When combining all pairs of location vectors are being associated and product of involved masses is assigned to the result set. Obtained assignment is supposed to increase the informative context of the initial structures. The combination of structures embracing measurement data is assumed to result in position fixing.

Position fixing can be achieved provided the association of sets enables the selection of common points located within intersection of introduced ranges. Adequate selection can be done with T-norm operations [16] used during association [17]. The simplest T-norm results in smaller values being taken from consecutive pairs of elements in associating vectors. This operation is used in numerical examples further presented in the chapter.

Two assignments can be combined with two-dimensional table. Simplified result of combination takes the form of the assignment presented in Eq. 6.7.

$$\mathrm{m}_c(e_c) = \{(\mu_{c1}(x_k), m_{c1}(\mu_{c1}(x_k))), .., (\mu_{cl}(x_k), m_{cl}(\mu_{cl}(x_k)))\} \qquad (6.7)$$

The result of association with T-norm operator may be empty or sub-normal. Therefore, a certain amount of mass is assigned to null set which means the conflicting situation that can also be referred to as an inconsistency. In position fixing inconsistency means the absence of a hypothesis space point within intersection area. This might indicate poor quality of considered set of measurements as well as scarce, consequently wrongly distributed, hypothesis points. Thus it is important to record all conflicting cases and evaluate its final uncorrupted value. It causes the achieved assignment Eq. 6.7 to become a pseudo belief structure, which is to be subject of transformation in order to obtain its normal version. The conversion introduced in this chapter is different; it yields an assignment with all normal fuzzy sets without taking into account the total sum of masses. The concept that lies behind this normalization is related to the attempt to avoid a corrupted probability being assigned to the fix. It is supposed that final credibility solely depends on probabilities related to selected ranges and initial weights attributed to each piece of evidence.

Before introducing this form of transformation let us concentrate on the exploration of combined assignment in order to make a fix. Table 6.2 presents results of combination of two belief structures shown in Table 6.1. Results of pair wise sets associations using the aforementioned T-norm along with obtained masses are shown.

Table 6.2 contains vectors which grades mean possibilities of hypothesis points fuzzy locations within regions of the intersection of involved ranges. Masses assigned to these vectors are products of probabilities attributed to appropriate ranges. It should be noted that these masses represent the credibility of a fix being located inside the intersection. In the probabilistic approach, the obtained result is a Bayesian representation of evidence [4, 18]. Grades can be perceived as probability of points being located within particular intersection. The total sum of grades within a single column is equal to one. Thus, any final probabilities supporting the hypothesis regarding representing the fix by lth point can be calculated with Eq. 6.8. The last row in Table 6.2 contains adequate set of data regarding each hypothesis frame point.

$$p(x_l) = \sum_{r=1}^{n} p(x_l \mid o_r) \cdot p(o_r) \tag{6.8}$$

where, for the example data, the following hold:
$o_r = o_{ij} = o_{i1} \cap o_{j2}$—intersection of ith and jth strips, note that the indexes refer to belief structures rows rather than to ranges established in the vicinity of an isoline $p(x_l \mid o_r)$—conditional probability that point x_l represents the fix provided intersection o_{ij} is considered
$p(o_r) = m_{cij}$—probability that the fix is located within intersection o_{ij}
$n = 4$—number of items in combined structure

Table 6.2 Results of combined location vectors and their masses

$\mu_{cij}(x_k)$[a]	x_1	x_2	x_3	x_4	m_{cij}[b]
$\mu_{c11}(x_k)$	{ 1	0	0	0 }	$m_{c11} = 0.06$
$\mu_{c12}(x_k)$	{ 0	0	1	0 }	$m_{c12} = 0.14$
$\mu_{c21}(x_k)$	{ 0	0.8	0	0 }	$m_{c21} = 0.24$
$\mu_{c22}(x_k)$	{ 0	0.2	0	1 }	$m_{c22} = 0.56$
$p(..)^c =$	{ 0.06	0.304	0.14	0.56 }	

[a] $\mu_{cij}(x_k)$—stands for non-normalized location vector being result of combination of ith and jth sets of two belief assignments
[b] m_{cij}—stands for the result mass assigned to ijth combined location vector
[c] $p(..)$—are final probabilities on representing the fix by each of the hypothesis space point

From a possibility standpoint, the result appears as pseudo belief structure (that is condition of which Eq. 6.4 is not supported). Before drawing final conclusion on the fix or undertaking further combination steps, the pseudo belief structure is to be converted to its normal state. There are two basic concepts regarding normalization. They are named after the people who proposed their first versions: it is usually said about Dempster and Yager [11] methods. The results of the Yager method application in order to normalize data from Table 6.2 are collated in Table 6.3. In the method all grades are increased by complement of their highest value. The result of modification of obtained data is seen in row $\mu^{Y}_{c21}(x_k)$ of the table. As a consequence of such modification, zero grades referring to empty locations gain some value. Therefore, the detection ability of inconsistency cases is impaired.

In the Dempster concept, masses assigned to non-empty sets are increased by factor which is calculated based on total inconsistency mass. Note that probability value attributed to the final fix increases in case of conflicting, erratic evidence.

Results of the Dempster method application to non-normalized data gathered in Table 6.2 are shown in Table 6.4. Superscripted D is used to distinguish sets of data from those obtained with the method proposed in the chapter.

Normalized data must be further examined in order to calculate probabilities of representing the fix by each hypothesis frame point. Support plausibility measure,

Table 6.3 Results of combined location vectors and their masses normalized with Yager method

$\mu^{Y}_{cij}(x_k)$[a]	x_1	x_2	x_3	x_4	m^{Y}_{cij} [b]
$\mu^{Y}_{c11}(x_k)$	{ 1	0	0	0 }	$m_{c11} = 0.06$
$\mu^{Y}_{c12}(x_k)$	{ 0	0	1	0 }	$m_{c12} = 0.14$
$\mu^{Y}_{c21}(x_k)$	{ 0.2	1	0.2	0.2 }	$m_{c21} = 0.24$
$\mu^{Y}_{c22}(x_k)$	{ 0	0.2	0	1 }	$m_{c22} = 0.56$
$pl^{Y}(..)$[c] =	{ 0.108	0.352	0.188	0.608 }	

[a] $\mu^{Y}_{cij}(x_k)$—stands for normalized, with Yager method, location vector being result of combination of *i*th and *j*th sets of two belief assignments
[b] m^{Y}_{cij}—stands for the result mass assigned to *ij*th normalized location vector
[c] $pl^{Y}(..)$—are final plausibility measures on representing the fix by each of the hypothesis space point

Table 6.4 Results of combination; associated location vectors and their masses normalized with two other methods

μ^{F}_{cij} [a] & μ^{D}_{cij} [b]	x_1	x_2	x_3	x_4	m^{F}_{cij} [a]	m^{D}_{cij} [b]
$\mu^{F}_{c11} = \mu^{D}_{c11}$	{ 1	0	0	0 }	0.06	0.063
$\mu^{F}_{c12} = \mu^{D}_{c12}$	{ 0	0	1	0 }	0.14	0.147
$\mu^{F}_{c21} = \mu^{D}_{c21}$	{ 0	1	0	0 }	0.24	0.202
$\mu^{F}_{c22} = \mu^{F}_{c22}$	{ 0	0.2	0	1 }	0.56	0.588
$pl^{F}(..)$[a] =	{ 0.060	0.304	0.140	0.560 }		
$pl^{D}(..)$[b] =	{ 0.063	0.319	0.147	0.588 }		

[a]—superscripted F denotes data obtained for transformation method proposed in the chapter
[b]—superscripted D denotes data obtained for normalization based on the Dempster approach

included in n related sets, that lth point represents the fix can be calculated using Eq. 6.9.

$$pl(x_l) = \sum_{k=1}^{n} \mu_k(x_l) \cdot m(\mu_k(x_i)) \qquad (6.9)$$

Where component $m(\mu_k(x_i))$ denotes credibility mass attributed to kth intersection involving ith and jth vectors. Factor $\mu_k(x_i)$ represent the abbreviated $\mu_{cij}(x_i)$ to reflect the fuzzy location of the lth hypothesis point within given ranges intersection.

Equation 6.9 calculates the plausibility measure of support for a certain fuzzy set embraced in family of related items, it was derived in [6]. Plausibility and belief are basic measures used in MTE [9] they represent limits of interval valued probability of support for selected hypothesis. It should be remembered that the same formula has been derived when Bayesian evidence representation is considered (see Eq. 6.8), also refer to [18] for example detailed analyses on Bayesian reasoning. The second last row in Table 6.4 embraces collection of data with title $pl^F(..)$, the set embraces data calculated with Eq. 6.9. The results were obtained using conversion method presented later in the chapter. Please also note that the probabilities calculated with Bayesian reasoning scheme are exactly the same.

In the possibilistic approach towards position fixing plausibility and belief measures are those that enable solution selection. Support belief for the lth hypothesis point embraced in certain family of sets can be calculated with Eq. 6.10.

$$bel(x_l) = \sum_{k=1}^{n} \min_{x_i \in \Omega_H;\, i \neq l} \neg\mu_k(x_i) \cdot m(\mu_k(x_i)) \qquad (6.10)$$

Equation 6.10 intended for position fixing is a reduced version of the general one included in [9]. Reduction is made due to special kind of the reference fuzzy set used in making the fix. To calculate belief value one has to find a minimum among the complemented grades ($\neg$) of each set, grade of interest is to be omitted. It should be noted that multiple point presence within given intersection of ranges causes the belief for each of points to be zeroed. For this reason, in presented form, belief cannot be considered as the primary factor in discussed position fixing problem. It can be useful while identifying the uniqueness of the solution.

Plausibility measure, meant as a crucial factor in position selection, solely depends on hypothesis point locations within selected areas and probabilities assigned to these areas. It is also assumed that normalization procedure should not affect the final fix.

Figure 6.4 presents the outputs generated by FIXMTE the implemented software. The illustration demonstrates fixed position made based on measurements of four distances. Estimated errors assigned to each isoline as well as their crisp subjective evaluations are shown in the insertion. It should noted that the term "line of position" instead of "isoline" is used in the picture description. In nautical science both of them are frequently used as synonyms.

Iterative procedure was used to make the fix. In consecutive iterations a decreasing search area is explored, grids covering search area during the first two iterations are

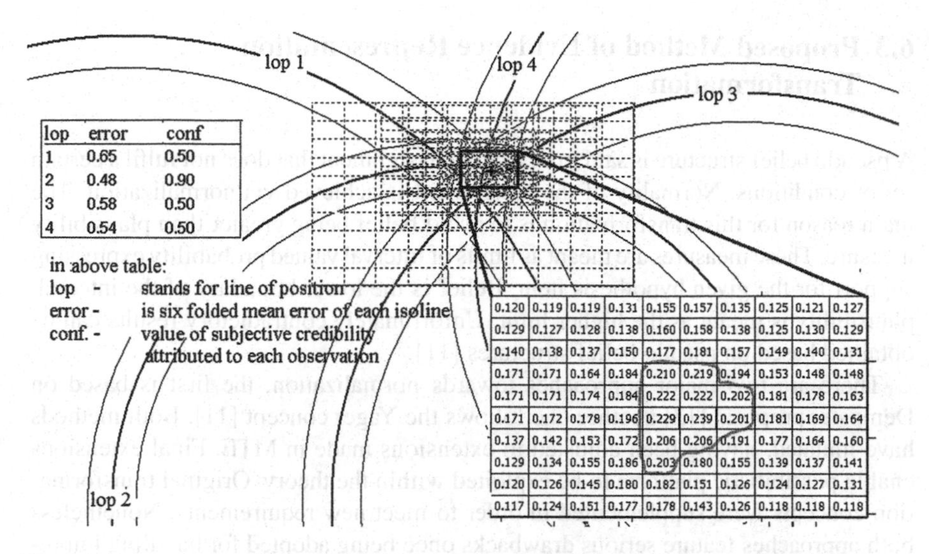

Fig. 6.4 Output generated by software implementing proposed approach to position fixing

clearly shown in the picture. It is assumed that, for certain computation stages, considered area embraces all maxima points selected during previous iterations. A mash of 10×10 points was spanned over the explored area.

In Fig. 6.4 grid embracing distribution of the fix plausibility measures all over examined area for last iteration is inserted. Small cross marks the point considered as fixed position, for which support plausibility measure is the highest one and equal to 0.239. Scheme of calculations embraces further presented algorithm. It should be noted that inconsistency masses do not increase uncertainty; therefore the support plausibility measure is relatively small.

This approach creates an opportunity to revise the fix quality evaluation. Traditional meaning of the fix accuracy is related to an area around the fixed position. Within the area the true position of the ship is located with a certain degree of credibility. It is often assumed that the area is of circular or elliptic shape, see Fig. 6.3. Formulas enabling calculation of its radius were derived for typical schemes followed while making a fix. Usually bearings and distances are taken. Two or three bearings combined with distances are often exploited for position fixing. Appropriate formula is to be engaged to a priori evaluating mean error of the fix [2].

In the proposed approach, distribution of probabilities of the fix being located within explored area is embedded into its methodology. Therefore, accuracy evaluation can be made a posteriori and is to be perceived as a cohesive area within which probability (plausibility) of the fix location is higher than the required threshold value (see exploded insert of Fig. 6.4).

6.3 Proposed Method of Evidence Representation Transformation

A pseudo belief structure is said to be a belief assignment that does not fulfil a certain set of conditions. Normality of the assignment is achieved via normalization. The main reason for this transformation is to avoid belief being greater than plausibility measure. These measures are meant as limits of interval valued probability expressing support for the given hypothesis item. Belief is the lower boundary of the interval, plausibility is meant as the higher limit. Unfortunately contradictory results can be obtained based on pseudo belief structures [11].

There are two major approaches towards normalization, the first is based on Dempster proposal [13] another one follows the Yager concept [11]. Both methods have mutated, having been adjusted to extensions made in MTE. Final extensions enable possibilistic platform to be exploited within the theory. Original transformation concepts were supplemented in order to meet new requirements. Nonetheless both approaches feature serious drawbacks once being adopted for handling imprecise measurements. In the Dempster concept, all masses assigned to non-empty sets, including those representing uncertainty, are increased by a factor that is a function of the total inconsistency mass. Final masses calculated based on initial assignments are increased during normalization with the modification factor. These lead to the confusing misbehaviour of the approach while handling low quality or contradictory evidence. One can coin unacceptable proposition that "the higher inconsistency mass the greater probability assigned to non empty sets" or referring to position fixing, "the poorer quality data, the higher credibility attributed to the fix". The Yager method's ability to detect cases of inconsistency is seriously impaired thus the quality of evidence at hand might be overestimated.

Herein modified approach towards modification of belief assignment is proposed. Result of conversion is not a belief structure since the sum of all involved masses is smaller than or equal to one. Apart from this all requirements (Eqs. 6.2–6.5) are observed; it also means that the converted assignment contains normal fuzzy sets.

The main advantage of the approach is the ability to maintain unchanged value of the plausibility measure, the primary factor deciding on selection of the final solution. It also assures that belief and plausibility measures remain in proper relation, the first is not greater than the second one.

In more formal language, the transformation can be expressed in following way:

- given pseudo belief structure of the form: $\mathrm{m}^* = \{(\mu_k^*, m^*(\mu_k^*(x_i)))\}$ with family of fuzzy sets $\{\mu_k^*(x_i)\}$ representing evidence and crisp set of masses $\{m^*(\mu_k^*(x_i))\}$;
- converted representation of evidence is sought: $\mathrm{m}^F = \{(\mu_k^F, m^*(\mu_k^F(x_i)))\}$ for which family of fuzzy sets $\{\mu_k^F(x_i)\}$ are normal, modified crisp set of masses $\{m^F(\mu_k^F(x_i))\}$ is assigned to new sets;
- for initial representation and its converted form, for each l, following condition is to be observed: $\sum_{k=1}^{n} \mu_k^*(x_l) \cdot m^*(\mu_k^*(x_i)) = \sum_{k=1}^{n'} \mu_k^F(x_l) \cdot m^F(\mu_k^F(x_i))$.

The last condition suppresses changes in the plausibility measure (see Eq. 6.9). The plausibility proved to be the most important factor in selecting position from the hypothesis frame. Transformation can be carried out with the following algorithm.

Algorithm 1.

1. for each kth fuzzy set calculate its height, select its greatest grade: $h_k = \max_{all\, i}(\mu_k^*(x_i))$
2. modify grades in non-empty kth set using the formula: $\mu_k^F(x_i) = \frac{\mu_k^*(x_i)}{h_k}$; $h_k \neq 0$
3. modify mass assigned to non-empty kth set using the formula: $m^F(\mu_k^F(x_i)) = m^*(\mu_k^*(x_i)) \cdot h_k$
4. calculate uncertainty mass: $m^F(\Omega) = m^*(\Omega)$
5. calculate total inconsistency mass: $m_u = \sum m(\emptyset)$
6. calculate total reduction of masses: $m_a = \sum_{k=1}^{n} m^*(\mu_k^*(x_i)) \cdot (1 - h_k)$

The result generated by the above algorithm is a pseudo normal belief structure since total of masses is smaller than one by sum of m_u and m_a values. Based on gathered experience it was proved that remaining stuck to belief structure as the only evidence representation is unjustified in considered field of application.

Thanks to modification made in step 2 all fuzzy and not empty sets are normal. It is required that the normal fuzzy set contains at least one grade equal to one. Note that scheme of null grades distribution remain intact, which is not the case in Yager transformation. Additional m_a ratio is assigned total reduction of masses. The value can be analysed during evaluation of the obtained solution quality.

Masses assigned to non empty sets are reduced in step 3 of the algorithm 1. It should be noted that grades normalization and reduction of masses leads to obvious equality: $\frac{\mu_k^*(x_l))}{h_k} \cdot m^*(\mu_k^*(x_i)) \cdot h_i = \mu_k^*(x_l) \cdot m^*(\mu_k^*(x_i)) = \mu_k^F(x_l) \cdot m^F(\mu_k^F(x_i))$, which causes last requirement in presented formal definition of the transformation to also be observed. It proved to be the most important feature of the conversion method.

Proposed conversion features the following properties:

1. masses attributed to location vectors are not subject to unjustified changes;
2. conflicts, as not zero masses assigned to null sets, increase additional factor and potentially uncertainty;
3. all fuzzy sets are normal, null grades remain unchanged, subsequently conflicts detection is not impaired;
4. plausibility value as a primary factor in selecting fixed position remain intact during conversion;
5. transformation maintain basic for MTE condition: $bel(x_i) \leq pl(x_i)$, stipulating that belief measure cannot exceed plausibility value.

The condition specified in point 5 is not straightforward and needs to be proven. Eqs. 6.9 and 6.10 used for plausibility and belief calculations are to be recalled in order to prove the proposition.

Table 6.5 Specification of terms used for plausibility and belief calculations

	in Eq. 6.9	as consequence in Eq. 6.10	Observed condition
1	$\mu_k(x_l) = 1$	$\forall_i \ \min_{x_i \in \Omega_H;\, i \neq l} \neg\mu_k(x_i) \leq 1$	$bel(x_l) \leq pl(x_l)$
2	$\mu_k(x_l) = 0$	$\exists_i \ \min_{x_i \in \Omega_H;\, i \neq l} \neg\mu_k(x_i) = 0$	$bel(x_l) \leq pl(x_l)$
3	$\mu_k(x_l) \in [0, 1]$	$\exists_i \ \min_{x_i \in \Omega_H;\, i \neq l} \neg\mu_k(x_i) = 0$	$bel(x_l) \leq pl(x_l)$

Table 6.5 presents the proof specified using three statements assumed for the same location vector $\mu_k^F(x_i)$ and its mass: $m(\mu_k^F(x_i))$. Three different factors, fragments in both expressions are considered and specified. In the second column of the specification all possible values of grades mentioned in Eq. 6.9 are shown. As the consequence of the particular value and normality of involved set, an appropriate fragment embraced in Eq. 6.10 must exist. Expressions describing such fragments are presented in third column. Note that in third column minima of grades complements of normal fuzzy sets are considered. Grade of interest, considered in column 1, is omitted (see also search condition in *min* operation inside Eq. 6.10).

In the presented specification all possible cases are exhausted therefore one can conclude that condition: $bel(x_l) \leq pl(x_l)$ remains valid for all converted belief assignments that contain all normal fuzzy sets.

6.4 Summary and Conclusions

Dealing with uncertain and imprecise evidence is a challenge in nautical science and practice. Formal descriptions of problems encountered in navigation involve models that accept imprecise, erroneous and therefore uncertain values. The concept should be followed regarding position fixing and its accuracy evaluation. It is the navigator who has to handle a set of random points delivered by various navigational aids from which he is supposed to indicate a point as being the position of his ship. Dispersions of points are governed by two dimensional approximate distributions. The fixed position is located somewhere in the vicinity of indications at hand. It is very similar, in case of measured distances, bearings or horizontal angles. The ship's position is located within the area of crossings of appropriate isolines that intersect inside the confined area.

Practical navigation is based upon probability theory. The basis is enough to define distributions of random variables that are assumed to be of measured value. It also enables a priori evaluation of fixes taken according to certain schemata since accuracy is calculated with formulas designated for selected schedules of observations taking into account the constellation of landmarks and approximate measurements error.

Models that include uncertainty can be created with MTE. The theory can be perceived as an extension of the Bayesian Concept. It also offers combination mechanism, enabling the enrichment of informative context of the initial evidence. Despite its broad ability, the theory still remains unpopular in the presented scope of interest. In the chapter the main emphasis was put on position fixing and its accuracy evaluation under imprecise evidence and uncertain knowledge. Knowledge regarding random error dispersions is imprecise in respect of types of distributions and their parameters. Knowledge also embraces the subjective assessment of each piece of available data. Uncertainty is usually expressed by the following statement: observation of clear, well pronounced objects should dominate in selection of the fix when compared to measurements taken of vague distant landmarks. In traditional approach there is no room for including all issues presented above. On quality evaluation, a navigator can reason a priori before taking measurements. The proposed method, apart from its flexibility, feature ability of thorough a posteriori analysis of the fix that is important issue in geodesy and navigation.

Measurement and indication data, along with nautical knowledge, can be encoded into belief functions [5]. Both knowledge and data are considered as evidence that is exploited in navigation. Belief functions in nautical applications represent evidence and are subject to combination in order to increase their informative context. Evidence representations and results of their combinations could be pseudo belief structures that are to be converted to avoid conflicting final results. Conflict arises when belief is greater than plausibility measure.

It is assumed that representations should be normalized at the initial and intermediate stages of processing, to avoid contradictory results. The most popular normalization procedures feature serious disadvantages therefore; the proposal of conversion has been submitted. Suggested transformation cannot be perceived as normalization since it does not yield a belief structure due to total mass that could be less than one. Most important is that its output contains normal fuzzy sets that proved to be enough to avoid basic conflict. Moreover, plausibility measures regarding the fix remains intact due to proposed conversion.

The result of structure combination is a two-dimensional table that embraces enriched data enabling reasoning on the fix and its accuracy. From the possibilistic viewpoint this result is a structure that contains distributions of possibilities regarding representing the fix by each point from hypothesis frame. Mechanisms and methods available in theory of evidence can be exploited in order to derive formulas for calculating interval valued probability. Interval value limits are equal to belief and plausibility measures.

Alternatively from a probabilistic standpoint, the obtained result of combination can be perceived as Bayesian evidence representation. It should be stressed that this standpoint is justified for limited range of selected cases, in general, the final structure does not fulfil probability distribution requirements, although in the presented example, it does. Doubtfulness, meant as uncertainty attributed to measurements, was deliberately removed from the initial representations of evidence. Disability of modelling uncertainty is said to be the main drawback of the probabilistic approach. Remaining unresolved dilemma as to whether the conditions stipulated in the probabilistic model are observed or not, one can exploit a Bayesian approach to reason on draft shape of formula for calculating support plausibility for "being a fix" by any point. It is not surprising that two approaches yield virtually the same final expressions; both of them are included in the chapter.

References

1. Burrus, N., Lesage, D.: Theory of evidence. Technical report no. 0307–07/07/03. Activity: CSI seminar, EPITA research and development laboratory, Cedex France, 08 Dec 2004
2. Jurdziński, M.: Principles of Marine Navigation. WAM, Gdynia (2008)
3. Filipowicz, W.: Belief structures in position fixing. In: Mikulski, J. (ed.) Transport Systems Telematics, Communications in Computer and Information Science 104, pp. 434–446. Springer, Heidelberg (2010)
4. Liu, W., Hughes, J.G., McTear, M.F.: Representing heuristic knowledge and propagating beliefs in Dempster-Shafer theory of evidence. In: Federizzi, M., Kacprzyk, J., Yager, R.R. (eds.) Advances in the Dempster-Shafer Theory of Evidence, pp. 441–471. Willey, New York (1992)
5. Filipowicz, W.: Evidence representation and reasoning in selected applications. In: Jdrzejowicz P., Ngoc Thanh Nguyen, Kiem Hoang (eds.) Lecture Notes in Artificial Intelligence, pp. 251–260. Springer, Heidelberg (2011)
6. Filipowicz, W.: New approach towards position fixing. Annu. Navig. **16**, 41–54 (2010)
7. Filipowicz, W.: Belief structures and their application in navigation. Methods Appl. Informatics **3**, 53–82 (2009)
8. Kłopotek, M.A.: Identification of belief structure in Dempster-Shafer theory. Found. Comput. Decis. Sci. **21**(1), 35–54 (1996)
9. Denoeux, T.: Modelling vague beliefs using fuzzy valued belief structures. Fuzzy Sets Syst. **116**, 167–199 (2000)
10. Klir, G.J., Parviz B.: Probability possibility transformations a comparison. Int. J. Gen Syst. **21**(1), 291–310 (1992)
11. Yager, R.: On the normalization of fuzzy belief structures. Int. J. Approximate Reasoning **14**, 127–153 (1996)
12. Lee, E.S., Zhu, Q.: Fuzzy and Evidence Reasoning. Physica-Verlag, Heidelberg (1995)
13. Dempster, A.P.: A generalization of Bayesian inference. J. Roy. Stat. Soc. B **30**, 205–247 (1968)
14. Shafer, G.: A Mathematical Theory of Evidence. Princeton University Press, Princeton (1976)
15. Yen, J.: Generalizing the DempsterShafer theory to fuzzy sets. IEEE Trans. Syst. Man Cybern. **20**(3), 559–570 (1990)
16. Rutkowski, L.: Methods and Techniques of the Artificial Intelligence. PWN, Warsaw (2009)
17. Yager, R.: Reasoning with conjunctive knowledge. Fuzzy Sets Syst. **28**, 69–83 (1988)
18. Liu, W., Hughes, J.G., McTear, M.F.: Representing Heuristic knowledge in D-S theory. In: Proceedings of the 8th Conference on Uncertainty in Artificial Intelligence (UAI'92), pp. 182–190, Morgan Kaufmann Publishers Inc., San Francisco (1992)

Chapter 7
Segmentation of Hyperspectral Images by Tuned Chromatic Watershed

Ramón Moreno and Manuel Graña

Abstract This work presents a segmentation method for multidimensional images, therefore it is valid for standard Red, Green, Blue (RGB) images, multi-spectral images or hyperspectral images. On the one hand it is based in a tuned version of watershed transform, and on the other hand it is based on a chromatic gradient that is made through Hyperspherical Coordinates. A remarkable feature of this algorithm is its robustness; it outperforms the natural oversegmentation induced by the standard watershed. Another important property of this algorithm is its robustness respect changes on the intensity: shines and shadows. Inspired on the Human Vision System (HVS) this algorithm provides segmentations according with the user expectations, where homogeneous chromatic regions of an image corespond with homogeneous convex regions of the output.

Keywords Hyperspherical Coordinates · Hyperspectral Chromatic Gradient · Watershed Transformation · Hyperspectral Image Segmentation · Computer Vision · Edge Detection

7.1 Introduction

Image segmentation is the main topic on computer vision and image procesing. It is one of the first steps in many applications and on its result depends the final process quality. Even though the idea of segmentation is easy to understand, it is an intrinsically difficult problem and it is very difficult to implement algorithms able to solve this problem, and of course, there are not algorithms for general purposes. That is due to some main reasons: First, segmentation is a human (or mamarian) hability,

---

R. Moreno (✉) · M. Graña
Computational Intelligence Group, Universidad del País Vasco, P Manuel Lardizabal 1, 20018 Donostia-San Sebastián, Spain
e-mail: ramon.moreno@ehu.es

J. W. Tweedale and L. C. Jain (eds.), *Recent Advances in Knowledge-based Paradigms and Applications*, Advances in Intelligent Systems and Computing 234,
DOI: 10.1007/978-3-319-01649-8_7,

therefore, different people can provide different segmentations of the same picture; and second, illumination changes have strong effects on scene perception, then a scenary with different illumination provide different captures. This work presents a segmentation algorithm for hyperspectral and multispectral images which cover both previous points and is the continuity of some previous research [1, 2].

Computational methods for image segmentation can be classified into three sets: supervised, semi-supervised and unsupervised methods. Supervised methods are those which require prior knowledge [3, 4], these methods need to know the true reality before learning rules. Unsupervised methods are those which do not need prior knowledge to carry out segmentation [5–7]. Finally semi-supervised methods are a mixture of both, generally they use statistical approaches [8, 9].

Edge detection in hyperspectral images is an intrinsically difficult problem. There are few strategies used for hyperspectral imagery processing, which can be subdivided on the basis of their principle procedures using two techniques: First monochromatic-based techniques treat information from the spectral bands or their combination individually, and then combine the individual results together [10]; and second, vector-valued techniques treat the spectral information as vectors in spectral space, for example, spectral angle mapper [11–13].

In accordance with this division of segmentation methods and edge detectors, this proposal is an unsupervised method that uses a gradient approach which belongs to vector-valued techniques.

Watershed transformation is an unsupervised method that is usually applied over the image gradient. It was introduced in image analysis by Beucher and Lantuejoul [14], and subsequently a lot of algorithm variations and applications have been proposed [10, 15, 16]. In short, it considers a bi-dimensional image as a topographic relief map. The value of a pixel is interpreted as its elevation. The watershed lines divide the image into catchment basins, so that each basin is associated with one local minimum in the topographic relief map. The watershed transformation works on the spatial gradient magnitude function of the image domain. The crest lines in the gradient magnitude image correspond to the edges of image objects. Therefore, the watershed transformation partitions the image into meaningful regions according to the gradient crest lines.

The key idea of watershed transformation is to apply the transformation over the image gradient. The gradient gives pixels of change in the image domain, therefore a suitable gradient could be used as an edge detector. This work proposes a well suited image gradient which belongs to the vector-valued set because it uses all spectral information. This gradient is made by Hyperspherical coordinates, where the use of Hyperspherical coordinates gives a physical meaning because it is related to the Extremadura Research Centre for Advanced Technologies (DRM) [17] and takes profit of it in order to avoid shines and shadows. On the other hand, watershed transformation output has oversegmentation, that is like a mosaic. The method presented on this chapter avoids this drawback thanks to a threshold and to the use of a Gaussian blurring over the image gradient. The method is summarized below:

(1) Transform the image to Hyperspherical coordinates

(2) Make the chromatic gradient
(3) Apply the t-Watershed method.

So as to demostrate the behavior of this proposal, the test of this segmentation method is done with some images taken in our lab with a SOC 710 hyperspectral camera.

This chapter is outlined as follows: Sect. 7.2 presents the Hyperspherical transform. Section 7.3 presents a chromatic gradient. Section 7.4 presents the segmentation method. Section 7.5 shows the experimental results and Sect. 7.6 finishes with the conclusions.

7.2 Hyperspherical Coordinates

This section illustrates the hyperspherical transformation. Hyperspheres are also known as n-spheres. It is the spherical expresion of a point when dimensionality (d) is bigger than 3. There are some specific names for some of these transfomations for example, when $d = 4$ is named quaternion, $d = 6$ octonion. Usually hyperspecrtal images have more than 30 bands–dimensions–therefore we refer it as Hyperspherical transformation. This transformation can be handle solely through the following way.

Let denote p a hyperspectral pixel color in n dimensional Euclidean space. In Cartesian coordinates it is represented by $p = \{v_1, v_2, v_3, ..., v_n\}$ where v_i is the coordinate value of the i-th dimension. This pixel can be represented in Hyperspherical coordinates $p = \{l, \phi_1, \phi_2, \phi_3, .., \phi_{n-1}\}$, where l is the vector magnitude that gives the radial distance, and $\{\phi_1, \phi_2, \phi_3, .., \phi_{n-1}\}$ are the angular parameters. This coordinate transformation is performed by the following expression described on Eq. 233.1.

There are some exceptions: if $v_i \neq 0$ for some i but all of $v_{i+1}, \cdots, v_n$ are zero then $\phi_i = 0$ when $v_i > 0$. When all $v_i, \cdots, v_n$ are zero then ϕ_i is undefined, usually a zero value is assigned. From Euclidean representation to Hyperspherical representation solely by using the set of Eqs. 233.1.

A more compact notation for the Hyperspherical Coordinates is $p = \{l, \bar{\phi}\}$, where $\bar{\phi}$ is the vector of size $n - 1$ containing the angular parameters. Given a hyperspectral image $I(x) = \left\{(v_1, v_2, v_3, ..., v_n)_x ; x \in \mathbb{N}^2\right\}$, where x refers to the pixel coordinates in the image domain, we denote the corresponding hyperspherical representation as $P(x) = \left\{(l, \bar{\phi})_x ; x \in \mathbb{N}^2\right\}$, from which we use $\bar{\phi}_x$ as the chromaticity representation of the pixel's and l_x as its (grayscale) intensity.

$$
\begin{aligned}
l &= \sqrt{v_1^2 + v_2^2 + v_3^2 + \cdots + v_n^2} \qquad (233.1)\\
\phi_1 &= \cot^{-1} \frac{v_1}{\sqrt{v_2^2 + v_3^2 + \cdots + v_n^2}}\\
\phi_2 &= \cot^{-1} \frac{v_2}{\sqrt{v_3^2 + v_4^2 + \cdots + v_n^2}}\\
&\vdots
\end{aligned}
$$

$$\phi_{n-2} = \cot^{-1} \frac{v_{n-2}}{\sqrt{v_{n-1}^2 + v_n^2}}$$

$$\phi_{n-1} = 2\cot^{-1} \frac{\sqrt{v_{n-1}^2 + v_n^2} - v_{n-1}}{v_n}$$

Figure 7.1 shows an illustrative example in order to clarify the meaning of the chromaticity in the hyperspectral image domain. It is a synthetic hyperspectral image of 5 × 5 pixels and 200 spectral bands. Each pixel spectral signature has the same Gaussian shaped profile but with different peak height, corresponding to different image intensity as can be appreciated in Fig. 7.1a showing the image intensity $\{l_x\}$. Figure. 7.1b shows the spectral signature of all pixels in the Cartesian coordinate representation, Figure 7.1c shows the chromatic spectral signature $\{\bar{\phi}_x\}$ which is the same plot for all pixels. The chromaticity $\bar{\phi}$ thus defines a line in the n-dimensional space of hyperspectral pixel colors of points that only vary their luminosity l.

According to the aforegoing coordinate transformation, we can perform the following hyperspectral separation. Given a hyperspectral image $I(x)$ in the traditional Cartesian coordinate representation it can computed the equivalently hyperspherical representation $P(x) = \{(l, \bar{\phi})_x; x \in \mathbb{N}^2\}$. Then, it is possible to construct the separate intensity image $L(x) = \{(l)_x; x \in \mathbb{N}^2\}$, and the chromaticity image $C(x) = \{(\bar{\phi})_x; x \in \mathbb{N}^2\}$. In the synthetic example shown at Fig. 7.1, $I(x)$ pixels are plotted in Fig. 7.1b, the spectral chromaticity $C(x)$ in Fig. 7.1c and the image intensity $L(x)$ in Fig. 7.1a. This separation allows us the independent processing of hyperspectral color and intensity information, so that segmentation algorithms showing color constancy can be defined in the hyperspectral domain. This decomposition can be also embedded in models of reflectance like the Extremadura Research Centre for Advanced Technologies (DRM) [17] or the Bidirectional Reflection Distribution Function (BRDF) [18] where they can be decomposed as diffuse and specular components.

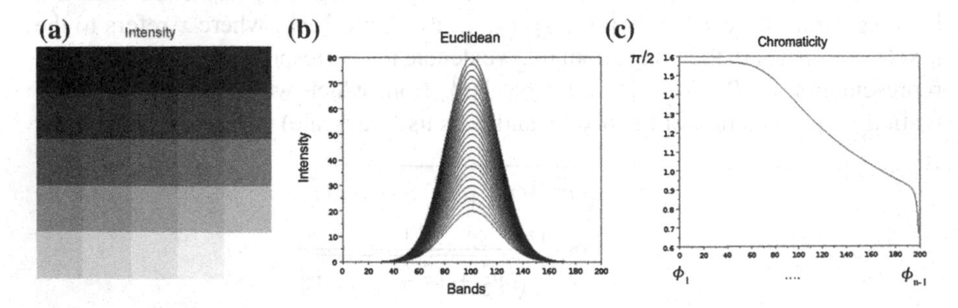

Fig. 7.1 Synthetic image (**a**) the image intensity $\{l_x\}$, (**b**) shows the Gaussian shaped signature profile of all the pixels, and (**c**) shows the angle components of the hyperspherical coordinates shared by the spectral signatures of all pixels in the image, corresponding to the common chromaticity of the pixels

7.3 Chromatic Gradient

Image gradient is a very useful tool on image processing, it measures the change between neighboring pixels. Formaly it is defined by the derivatives and Laplacian transformations. Figure 7.2 shows the graphic expression of the image gradient. Region with homogeneous color have low gradient—see vector longitude in the background— whereas regions with bigger changes have bigger grandient too–see the vectors' longitude centre of the image—Those vectors have two components: direction and magnitude. Usually in computer vision only the magnitude is used. In this work uses only the gradient magnitude, thus, gradient image output is a bi-dimensional image of numbers, where pixels with high intensity means that this pixel is in a region of change.

In order to avoid the effect of illumination changes (shines and shadows) this work proposes a chromatic gradient. This gradient is independent of the intensity therefore it is less sensitive to changes produced by the illumination. The Chromatic gradient is deffined as follows.

Linear convolution gradient operators, such as the Prewitt operators with the underlying topology is the one induced by the Euclidean distance defined on the Cartesian coordinate representation [19]. In order to define a chromatic gradient operator, we may assume a kind of non-linear convolution where the convolution mask has the same structure as the Prewitt operators, but the underlying chromatic distance is based only on the chromaticity as follows: For two pixels p and q it is computed the Manhattan or Taxicab distance [20] on the chromatic representation of the pixels:

$$\angle(p,q) = \sum_{i=1}^{n-1} \left|\bar{\phi}_{p,i} - \bar{\phi}_{q,i}\right| \tag{7.1}$$

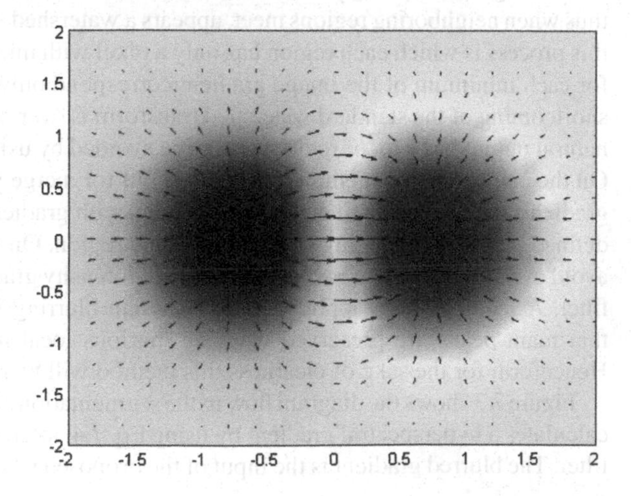

Fig. 7.2 Image gradient function

Note that the chromatic distance defined in Eq. 7.1 is always positive. Note also that the process is non linear, so it can not expressed it by linear convolution kernels. The row pseudo-convolution operator is defined as:

$$CG_R\left(\mathrm{C}\left(i,j\right)\right)=\sum_{r=-1}^{1}\angle\left(\mathrm{C}\left(i-r,j+1\right),\mathrm{C}\left(i-r,j-1\right)\right),\tag{7.2}$$

and the column pseudo-convolution is defined as:

$$CG_C\left(\mathrm{C}\left(i,j\right)\right)=\sum_{c=-1}^{1}\angle\left(\mathrm{C}\left(i+1,j-c\right),\mathrm{C}\left(i-1,j-c\right)\right),\tag{7.3}$$

so that the color distance between pixels substitutes the intensity subtraction of the Prewitt linear operator. By using Eqs. 7.2 and 7.3 the hyperspectral chromatic gradient magnitude image is computed as:

$$CG(x)=CG_R\left(x\right)+CG_C\left(x\right)\tag{7.4}$$

7.4 Segmentation Method

This segmentation method is based on watershed which it is applied into hyperspectral images following a straightforward way. This work advocates the suitability of the aforegoing hyperspectral gradient in order to detect the true edges avoiding the effects of intensity changes; shines and shadows.

Watershed transform uses a gradient image who interpret as a topographic relief. Then it begins a flooding process which it is carried out from each minimal gradient, thus when neighboring regions meet, appears a watershed edge. The natural output of this process is which each region has only a pixel with minimum gradient, therefore, for each minimum of the image gradient correspond only a region. Consequently a shortcoming of the standard watershed transform is over-segmentation. Thereby, the natural output has a lot of regions. It can be avoided by using two strategies together. On the one hand, this method uses a threshold for merge neighbor regions with low gradient; in other words, neighboring regions with gradient intensity lower than the defined threshold are going to be merged in a region. On the other hand, in order to avoid over-segmentation in regions with high intensity gradient, it applies a Gaussian filter. A feature of the application of Gaussian blurring into the gradient image is that main peaks are preserved whereas spurious local minimum can be avoided. Henceforth for the sake of clearness this method will be refered as ‘t-Watershed’.

Figure 7.3 shows the diagram flow of the segmentation algorithm. First, the method calculates a hyperspectral gradient by using Eq. 7.4, afterwards it applies a Gaussian filter. The blurred gradient is the input of the proposed t-Watershed transform. At the

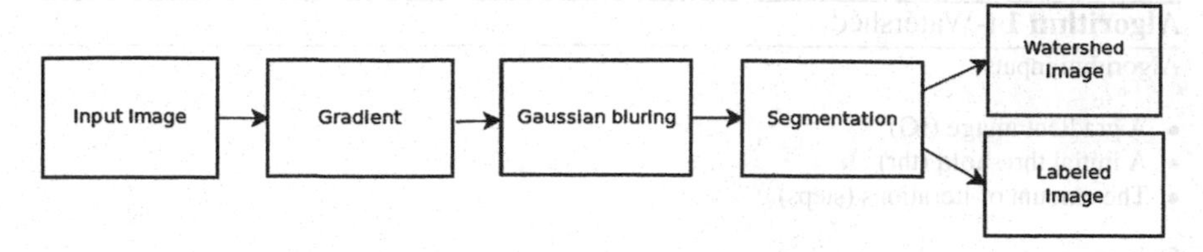

Fig. 7.3 Hyperspectral segmentation diagram flow

end, the method output gives an image with the watershed edges and other one with the labeled regions.

7.4.1 t-Watershed Algorithm

t-Watershed transform has the advantage that thanks to the parameter *threshold* it can control the beginning of the flooding process. In other words, each convex region in the image with a gradient intensity lower than *threshold* is going to be a region label. Algorithm 1 shows the pseudo-code of the method. The output of this algorithm is an image with the watershed edges and other one with the region labels.

7.5 Experimental Results

The proposed algorithm has been tested over the hyperspectral images taken with a SOC 710 camera. Main goal of this segmentation algorithm is to provide robustness regarding illumination effects, as shines, shadows and sudden illumination changes, looking for the true surface features. This one is the reason why it applies Hyperspherical Coordinates, which for the segmentation process are included into the chromatic gradient. In these experiments shows how robust is the proposed method respect to illumination changes. For all experiments we use a $thr = 0.2$ and $steps = 100$.

Images provides by the SOC 710 camera has 128 bands from 400 nm to 1000 nm, albeit that first and last bands are very noised, therefore for these experiments the seventy middle bands from 20th band to 90th band has been selected, thereby the spectral used range is from 493 nm to 728 nm which corresponds with the main visible light spectra.

Experiments are done on the images who have been taken with the SOC 710 camera; a blue ball and the synthetic orange. This image has some interesting properties. Both have an homogeneous gree background, and both have an spot and a shadow. On the other hand the blue ball has a smooth surface whereas the orange surface is wrinkled. Remainding the goal of this work, the aim is detect the true chromatic edges avoiding the intensity changes.

Algorithm 1 t-Watershed

Algorithm inputs :

- A gradient image (IG)
- A initial threshold (thr)
- The amount of iterations (steps)

Outputs:

- The watershed image (WS)
- An image with the labeled regions (IL).

Begin

1. First, it initializes the output images, and defines the intensity jump for thresholding into the image gradient.
2. By using a threshold, it finds the regions with minimal gradient, and helped by the primitive 'bwlabel' initializes the image of labels.
3. Then the algorithm begins with the flooding process who is going to finish after '**steps**' iterations.

   whereas iterations++ < steps ***loop***

   a. It calculates the new threshold (**thi**), and using it on the image gradient, finds the new pixels to label
   b. For each pixel who is not labeled already, it finds into the respective neighborhood if some of them has a label
   c. Depending of the labels found into the neighborhood, algorithm does different things:
      i If there is not labels, it creates a new label and assigns it to the current pixel
      ii If there is only a label, it assigns it to the current pixel
      iv If there are several labels
         A. If the gradient intensity is lower than the parameter '**thr**', merges all regions in a label and assigns it to the current pixel
         B. In other case, it marks it as watershed pixel

   end loop

end

Figure 7.4 shows the results. First row of images shows the process for the blue ball whereas the second one shows the synthetic orange output process. Column (a) are the intensity images. Column (b) are chromatic gradient. Column (c) shows the burred output of the detected edge. And column (d) shows the output of the proposed t-Watershed algorithm, where we can appreciate, which the segmentation gives the correct segmentation avoiding the shines and the shadows.

In order to compar this result with traditional intensity gradient, it is computed the same algorithm with the the Euclidean coordinates (the traditional representation). Figure. 7.5 shows the results. Like Fig. 7.4 first row of images is the smooth blue ball and bottom row is the srinked orange. By columns, (a) are the original intensity images, (b) shows the intensity gradient and (c) shows the blured gradient. Finally, column (d) shows the t-watershed segmentation. The parameters of this experiement are: $thr = 0.02$ on the blue ball and $thr = 0.1$ on the orange. As we can see, by

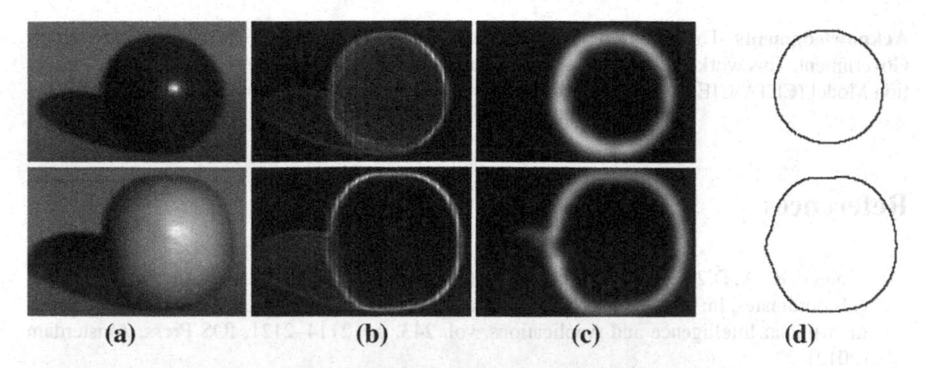

Fig. 7.4 Chromatic t-watershed segmentation of images taken with SOC 710 camera

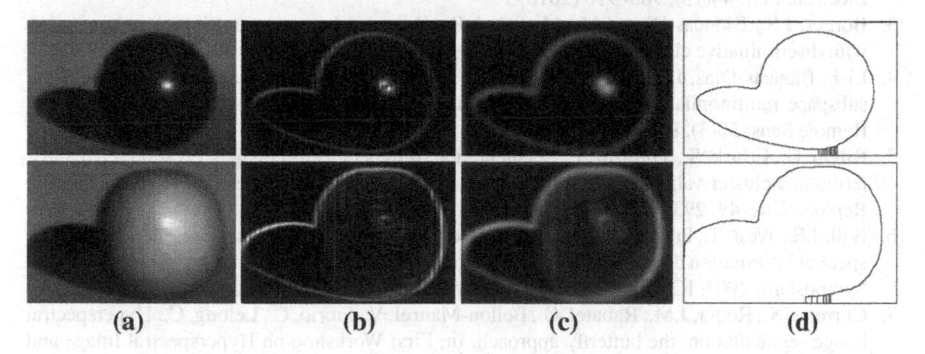

Fig. 7.5 Intensity t-watershed segmentation of images taken with SOC 710 camera

tuning thr parameter the algorithm is able to avoid the spot, however, it can not detect the ball without its shadow.

Clearly, the chromatic segmentation algorithm is able to detect the chromatic difference whereas using the traditional intensity gradient the algorithm can not detect this difference (due to shadowed regions).

7.6 Conclusion

This chapter proposes an improved version of the watershed transformation which provides good segmentations so as to avoid some well-known problems: the effect of changes on illumination and the standard watershed oversegmentation. This work has used two natural images taken with the SOC 710 camera. Both images have higlights and shadows, and the presented method shows well suited results.

Future work will use this proposed segmentation for color image clustering in order to stimate the amount of classes in a image.

Acknowledgments This work has been done thanks to the grant BFI08.271 of the Basque Country Government. This work was partially supported by the computing facilities of Dichromatic Reflection Model (CETA-CIEMAT), funded by the European Regional Development Fund (ERDF).

References

1. Moreno, R., A. D'Anjou.: Hyperspectral image segmentation by t-watershed and hyperspherical coordinates. In: Graa, M., Toro, C., Posada, J., Howlett, R.J., Jain, L.C. (eds.) KES. Frontiers in Artificial Intelligence and Applications, vol. 243, pp. 2114–2121, IOS Press, Amsterdam (2012)
2. Moreno, R., Graa, M., Zulueta.: RGB colour gradient following colour constancy preservation. Electron. Lett. **46**(13), 908–910 (2010)
3. Borges, J.S., Bioucas-Dias, J.M., Marcal, A.R.: Bayesian hyperspectral image segmentation with discriminative class learning. IEEE Trans. Geosci. Remote Sens. **49**, 2151–2164 (2011)
4. Li J., Bioucas-Dias, J.M., Plaza, A.: Spectral-spatial hyperspectral image segmentation using subspace multinomial logistic regression and markov random fields. IEEE Trans. Geosci. Remote Sens. **50**(3), 809–823 (2012)
5. Bilgin, G., Erturk, S., Yildirim, T.: Segmentation of hyperspectral images via subtractive clustering and cluster validation using One-Class support vector machines. IEEE Trans. Geosci. Remote Sens. **49**, 2936–2944 (2011)
6. Ball, J.E., West, T., Prasad, S., Bruce, L.M.: Level set hyperspectral image segmentation using spectral information divergence-based best band selection. In: Geoscience and Remote Sensing Symposium, 2007. IGARSS 2007. IEEE International, IEEE, 4053–4056 July 2007
7. Gorretta, N., Roger, J.M., Rabatel, G., Bellon-Maurel, V., Fiorio, C., Lelong, C.: Hyperspectral image segmentation: the butterfly approach. In: First Workshop on Hyperspectral Image and Signal Processing: Evolution in Remote Sensing, 2009. WHISPERS '09, IEEE, pp. 1–4, Aug 2009
8. Li, J., Bioucas-Dias, J.M., Plaza, A.: Semi-supervised hyperspectral image segmentation. In: First Workshop on Hyperspectral Image and Signal Processing: Evolution in Remote Sensing, 2009. WHISPERS '09, IEEE, pp. 1–4, Aug 2009
9. Li, J., Bioucas-Dias, J.M., Plaza, A.: Semisupervised hyperspectral image segmentation using multinomial logistic regression with active learning. IEEE Trans. Geosci. Remote Sens. **48**, 4085–4098 (2010)
10. Tarabalka, Y., Chanussot, J., Benediktsson, J.A.: Segmentation and classification of hyperspectral images using watershed transformation. Pattern Recogn. **43**(7), 2367–2379 (2010)
11. Dinh, V., Leitner, R., Paclik, P., and Duin, R.: A clustering based method for edge detection in hyperspectral images. In: Salberg, A.-B., Hardeberg, J., Jenssen R. (eds.) Image Analysis, Lecture Notes in Computer Science. vol. 5575, pp. 580–587. Springer, Heidelberg (2009)
12. Lee, M.A., Bruce, L.M.: Applying cellular automata to hyperspectral edge detection. .In: Geoscience and Remote Sensing Symposium (IGARSS), 2010 IEEE International, IEEE, pp. 2202–2205, July 2010
13. Luo, W., Zhong, L.: Spectral similarity measure edge detection algorithm in hyperspectral image. In: 2nd International Congress on Image and Signal Processing, 2009. CISP '09, IEEE, pp. 1–4, Oct 2009
14. Beucher, S., Lantuejoul, C.: Use of watersheds in contour detection. In: International Workshop on Image Processing: Real-time Edge and Motion Detection/Estimation, Rennes. France, Sept 1979
15. Elwaseif, M., Slater, L.: Quantifying tomb geometries in resistivity images using watershed algorithms. J. Archaeol. Sci. **37**(7), 1424–1436 (2010)
16. Dagher, I., Tom, K.E.: Waterballoons: A hybrid watershed balloon snake segmentation. Image Vis. Comput. **26**(7), 905–912 (2008)

17. Shafer, S.A.: Using color to separate reflection components. Color Res. appl. **10**, 43–51 1984
18. Hapke, B.: Bidirectional reflectance spectroscopy.1. theory. J. Geophys. Res. **86**, 3039–3054 (1981)
19. Yang, L., Zhao, D., Wu, X., Li, H., Zhai, J.: An improved prewitt algorithm for edge detection based on noised image. In: Image and Signal Processing (CISP), 2011 4th International Congress on, vol. 3, pp. 1197–1200, Oct 2011
20. Lavoie, T., Merlo, E.: An accurate estimation of the levenshtein distance using metric trees and manhattan distance. In: Software Clones (IWSC), 2012 6th International Workshop on, pp. 1–7, June 2012

Chapter 8
Impact of Migration Topologies on Performance of Teams of Agents

Piotr Jędrzejowicz and Izabela Wierzbowska

Abstract The chapter sums up the impact of some parameters defining the process of migration of data between asynchronous team of agents (A-Team) working in parallel in the architecture designed for solving difficult combinatorial optimization problems. A-Teams cooperate through exchange of intermediary computation results. The process of forwarding a result from one A-Team to another is called migration. Several known migration models have been compared, with different topologies and frequencies of migration. Also, an original model of communication has been proposed. The model, called *Randomized*, has no predefined migration topology, each A-Team sends data to another A-Team that is chosen at random. In this model migration is non-periodic and triggered only after an A-Team failed to improve its best current solution within a predefined time. All considered models, differing in migration topologies and frequencies, have been tested on instances of the Euclidean planar traveling salesman problem. The proposed model, *Randomized*, outperforms all other models under investigation, producing significantly better results.

Keywords A-Teams · Optimization · Computationally Hard Problems · Migration · TSP

8.1 Introduction

As it has been observed in Barbucha et al. [1] the techniques used to solve difficult combinatorial optimization problems have evolved from constructive algorithms to local search techniques, and finally to population-based algorithms. Technological

P. Jędrzejowicz (✉) · I. Wierzbowska
Department of Information Systems, Gdynia Maritime University, Morska 83, 81-225 Gdynia, Poland
e-mail: pj@am.gdynia.pl

I. Wierzbowska
e-mail: iza@am.gdynia.pl

J. W. Tweedale and L. C. Jain (eds.), *Recent Advances in Knowledge-based Paradigms and Applications*, Advances in Intelligent Systems and Computing 234,
DOI: 10.1007/978-3-319-01649-8_8,

advances have enabled development of various parallel and distributed versions of the population based methods. At the same time, as a result of convergence of many technologies within computer science, such as object-oriented programming, distributed computing and artificial life, the agent technology has emerged. An agent is understood here as any piece of software that is designed to use intelligence to automatically carry out an assigned task, mainly retrieving, processing and delivering information.

Paradigms of the population-based methods and Multiple Agent Systems (MAS) [19] have been during mid nineties integrated within the concept of the asynchronous team of agents (A-Team), a multi agent architecture, which has been proposed in several papers of Talukdar et al. [13–16].

The environments supporting implementation of A-Teams are represented by the JADE-Based A-Team environment (JABAT), built with the use of Java Agent Development Framework (JADE),[1] a framework proposed by TILAB [5]. Its subsequent versions and extensions were proposed in [2, 7] and [9]. JABAT complies with the requirements of the next generation A-Teams which are portable, scalable and in conformity with the standards provided by Foundation of Intelligent Physical Agents(FIPA).[2]

To solve a single task (a single instance of the problem being solved) JABAT uses a population of solutions that are improved by optimizing agents which represent different optimization algorithms. The agents work independently, in parallel and only cooperate indirectly using a common memory containing population of solutions.

In Jędrzejowicz and Wierzbowska [10] JABAT environment has been extended through integrating the team of asynchronous agent paradigm with the island-based genetic algorithm concept first introduced in Cohoon [6]. In Team of A-Teams (TA-Teams) communication or information exchange between cooperating A-Teams (islands) has been introduced. In, for example, Barbucha et al. [3] it has been shown that using TA-Teams a noticeable improvement in the quality of the computation results can be achieved.

A-Teams cooperate through exchange of intermediary computation results in a process called migration. The impact of the topology of migration on results in other models was considered in, for example, Rucinski [12], where the Island Model was considered. This chapter investigates how the choice of the migration topology and frequency influences results obtained by the TA-Teams solving instances of the Euclidean planar traveling salesman problem. The chapter is extended version of Jędrzejowicz and Wierzbowska [11].

The chapter is constructed as follows: Sect. 8.2 describes TA-Teams and the process of migration. Section 8.3 gives details of the experiments settings. Section 8.4 contains results of the carried out experiment. Finally, in Sect. 8.5, some conclusions and suggestions for future research are drawn.

---

[1] See http://jade.tilab.com/

[2] See http://www.fipa.org/

8.2 TA-Teams: Concept, Implementation and Settings

JABAT environment can be used to implement A-Team producing solutions to optimization problems using a set of optimizing agents, each representing an improvement algorithm. Such an algorithm receives one of the current solutions kept in the A-Team common memory, and attempts to improve it. Afterwards, successful or not, the result is send back to where it came from. The process of solving a single task (that is an instance of the problem at hand) consists of several steps. At first the initial population of solutions is generated and stored in the common memory. Individuals forming the initial population are, at the following computation stages, improved by independently acting agents (called optimization agents), each executing an improvement algorithm, usually problem dependent. Different improvement algorithms executed by different agents supposedly increase chances for reaching the global optimum. After a number of reading, improving and storing back cycles, when the stopping criterion is met, the best solution in the population is taken as the final result. The process is supervised by the agent called *SolutionManager*.

A typical JABAT implementation allows for running a number of A-Team in parallel providing the required computational resources are available. The teams, created and destroyed by so called *TaskManager*, never communicate and produce results independently (Fig. 8.1).

The JABAT implementation of TA-Teams allows for a number of A-Team to solve the same task by exploring different regions of the search space, with the added process of communication, that makes it possible to exchange best solutions between common memories maintained by each of the A-Team with a view to prevent premature convergence and assure diversity of individuals. A-Team with the added communication are called islands. Similar idea of carrying out the evolutionary process within subpopulations before migrating some individuals to other islands and then continuing the process in cycles involving evolutionary processes and migrations was previously used in, for example, Tanese [17] or Whitley et al. [18].

In TA-Teams the process of communication between common memories is supervised by a specialized agent called *MigrationManager* and defined by a number of parameters including:

- **Migration size:** number of individuals sent between common memories of A-Team in a single cycle;
- **Migration frequency:** length of time between migrations;
- **Migration topology:** an architecture defining which A-Team receives communication from another A-Team and sends communication to some other A-Team; and
- **Migration policy:** a rule determining how the received solution is incorporated into the common memory of the receiving A-Team.

The migration model used in TA-Teams is asynchronous. With a given frequency *MigrationManager* sends messages to islands, pointing out to which islands current

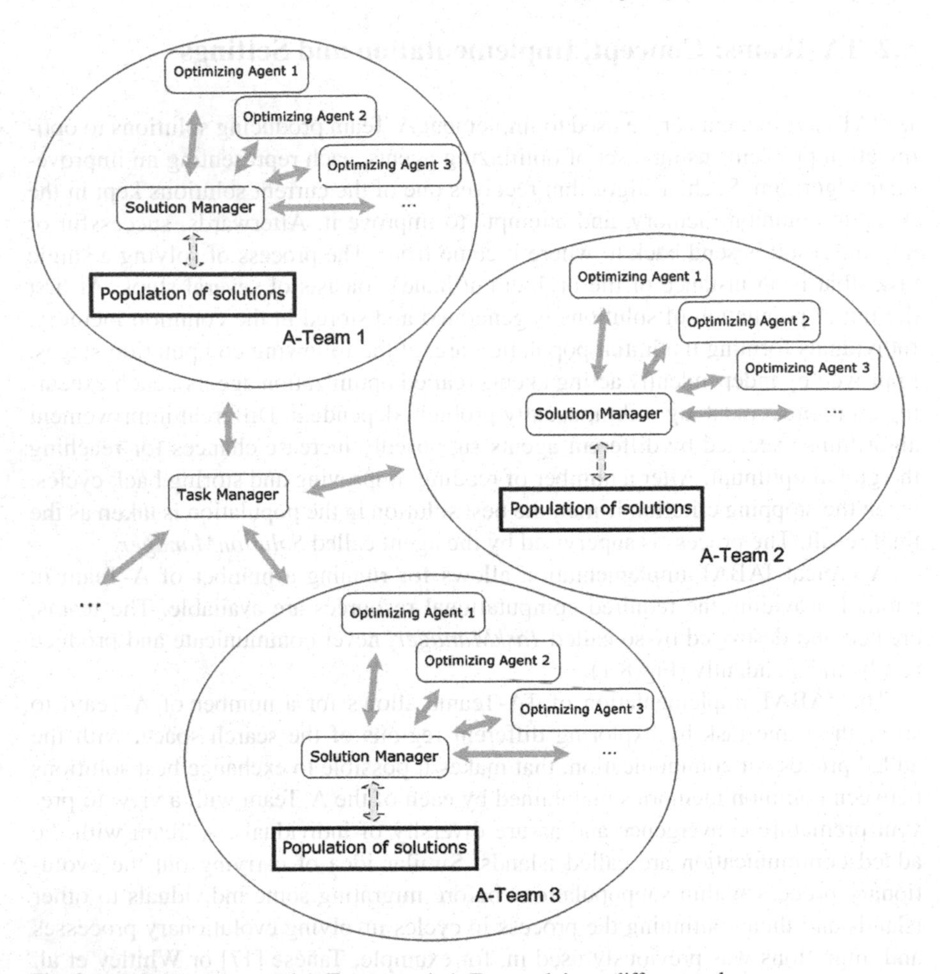

Fig. 8.1 JABAT with several A-Teams, each A-Team solving a different task

best solution should be send to. Then each A-Team (that is an island) receiving *MigrationManager* message, after reading it, sends current best solution to indicated island or islands. Figure 8.2 depicts the messages of the *MigrationManager* and how the solutions are sent between islands in the case of the *One Way Ring* topology.

8.2.1 Working Strategy

The process of solving a single task by an A-Team is controlled by the, so called, working strategy understood as a set of rules applicable to managing and maintaining the common memory, which contains a population of solutions called individuals.

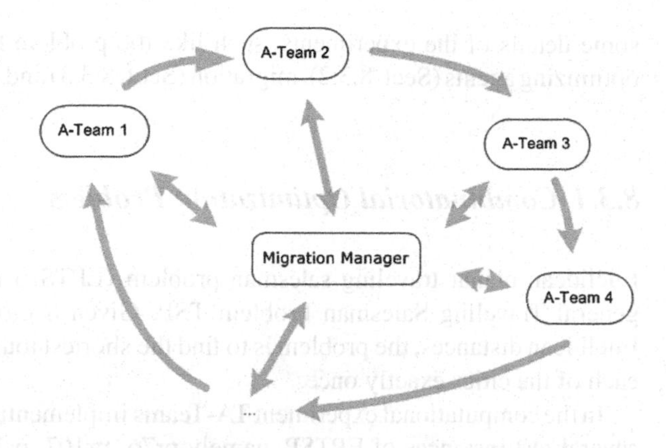

Fig. 8.2 A-Teams cyclically exchanging some solutions in the *One Way Ring* topology

A-Teams in TA-Teams follow the working strategy known as RB-RE (Random move selection with Blocking + Replacement of the worse with Exchange) which in Barbucha et al. [1] was identified as generating good quality solutions. In this strategy:

- All individuals in the initial population of solutions are generated randomly, the individuals are feasible solutions of the instance to be solved.
- Selection of individuals for improvement is a random move, however once selected individual (or individuals) cannot be selected again until all other individuals have been tried.
- Returning individual replaces the first found worse individual. If a worse individual cannot be found within a certain number of reviews (review is understood as a search for the worse individual after an improved solution is returned) then the worst individual in the common memory is replaced by the randomly generated one, representing a feasible solution. The number of reviews after which a random solution is generated equals 5.
- The computation time of a single A-Team is defined by the *no improvement time gap* = 2 min. If in this time gap no improvement of the current best solution has been achieved, the A-Team stops computations. Then all other A-Teams solving the same task stop as well, regardless of recent improvements in their best solutions.

The overall best result from common memories of all A-Teams is taken as the final solution found for the task.

8.3 Computational Experiment Design

The impact of applying different migration topologies and frequencies on the performance of the TA-Teams has been assessed by comparing results obtained from solving a well known combinatorial optimization problem. Section 8.3 describes some details of the experiments, such like the problem boing solved (Sect. 8.3.1), optimizing agents (Sect. 8.3.2), migration (Sect. 8.3.3) and other settings (Sect. 8.3.4).

8.3.1 Combinatorial Optimization Problem

Euclidean planar traveling salesman problem (EPTSP) is a particular case of the general Traveling Salesman Problem TSP. Given n cities in the plane and their Euclidean distances, the problem is to find the shortest tour, i.e. a closed path visiting each of the cities exactly once.

In the computational experiment TA-Teams implementation has been used to solve several test instances of EPTSP, namely pr76, pr107, pr144, pr299 and pr439 (the numbers in instances names indicate numbers of cities). The instances have been taken from well-known benchmark library for this problem, Traveling Salesman Problem Library (TSPLIB).[3]

8.3.2 Optimizing Agents

To solve instances of EPTSP the following optimization algorithms have been used as the inner algorithms of optimizing agents implemented within the system:

Simple exchange: deletes two random edges from the input solution thus breaking the tour under improvement into two disconnected paths and reconnects them in the other possible way, reversing one path.

Triple exchange: deletes three random edges, reconnects the tour in two other possible ways.

Recombination: there are two input solutions. A subpath from one of them is randomly selected. In the next step it is supplemented with edges from the second solution. If this happens to be impossible to add an edge, as the node has already been used in the subpath, the procedure constructs an edge connecting endpoint of the subpath with the closest point in the second input solution not yet in the resulting tour.

Mutation: two randomly selected points from the input solution are directly connected. This subpath is supplemented with edges from the input solution, as in *Recombination*.

Each of the optimizing agents chooses the best of the tours constructed by its inner algorithm in a number of trials. In each A-Team a set of four optimizing agents has been used, one of each kind.

[3] See http://www.iwr.uni-heidelberg.de/groups/comopt/software/TSPLIB95/

8.3.3 Migration Settings

In the experiment two migration parameters have been considered: *migration frequency* and *migration topology*. The values of *migration frequency* that have been used in the experiment were: 0.2, 0.6, 1, 1.4, 1.8, 2.2 and 2.6 min. As for the *migration topology* the following topologies have been considered (some of which are presented in Fig. 8.3):

One Way Ring: each A-Team receives communication from one adjacent A-Team and sends communication to another adjacent A-Team, the only unidirectional topology allowed;
Ring: two-directional ring;
Ring12 and Ring123: ring with additional communication between every second and every third node;
Broadcast: star topology;
Torus: rectangular lattice;
Lattice: of the size 2x4 or 4x4; and
Fully Connected: topology with the highest number of communication between islands.

Additionally a migration model called *Randomized* has also been considered, with the overhead caused by the information flow between islands reduced to minimum. Whenever an island needs a new solution, it sends appropriate message to *MigrationManager* and then receives the current best solution from another island, chosen at random by *MigrationManager*. An island asks for a new solution when the current best solution has not changed in half of the *no improvement time gap*.

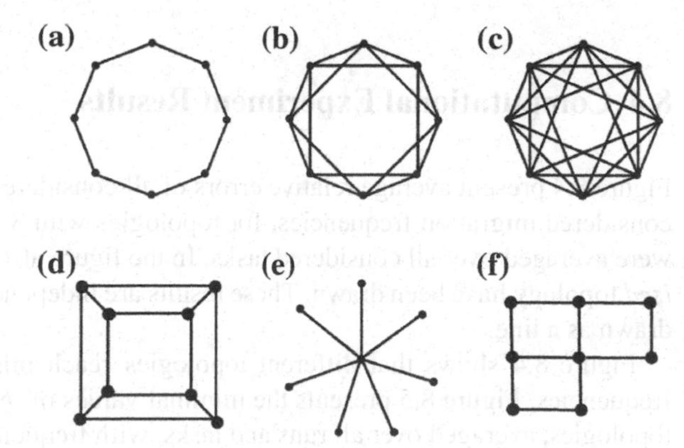

Fig. 8.3 Topologies drawn for 8 islands **(a)** *Ring* **(b)** *Ring12* **(c)** *Ring123* **(d)** *Broadcast* **(e)** *Torus* **(f)** *Lattice*

Two common migration settings are used in all models considered in this chapter:

Migration size = 1: in one cycle one current best solution is sent from the common memory of an A-Team to the common memory of another A-Team.

Best-worst migration policy: the current best solution is taken from the source population, the solution replaces the current worst solution in the target population.

8.3.4 *Other Settings*

In the experiment, topologies with 8 and 16 islands have been used. Each island shared a common memory consisting of 8 solution. Since for each island four optimizing agents have been working, the total number of optimizing agents used was $4*8 = 32$ and $4 * 16 = 64$, respectively.

The experiment has been carried out on the cluster Holk of the Tricity Academic Computer Network.[4] TA-Teams application has been implemented using JABAT, derived using JADE. As a consequence it has been possible to create so called containers on different machines and to connect them to the main platform. Agents may migrate from the main platform to these containers. Each instance used in the reported experiment was solved using five nodes on the cluster—one for the main platform and four to which the optimising agents migrate.

For all runs of each task and each pair of migration settings (frequency and topology) computation errors have been calculated in relation to the best results known for the problems. The results—in terms of relative computation error—have been averaged.

8.4 Computational Experiment Results

Figure 8.4 present average relative errors of all considered topologies for respective considered migration frequencies, for topologies with 8 and 16 islands. The errors were averaged over all considered tasks. In the figure also the results from *Randomized* topology have been drawn. These results are independent of the frequencies and drawn as a line.

Figure 8.4 shows that different topologies reach minimal values for different frequencies. Figure 8.5 presents the minimal values of relative errors for respective topologies, averaged over all runs and tasks, with frequencies for which the minima have been obtained, for respective numbers of islands. The results for *Randomized*

[4] Holk: 256 Intel Itanium 2 Dual Core with 12 MB L3cache processors, 2,3 TB total system memory, 4 TB disk storage. Mellanox InfiniBand interconnect with 10 Gb/s bandwidth. See http://www.task.gda.pl/english/hpc

(a)

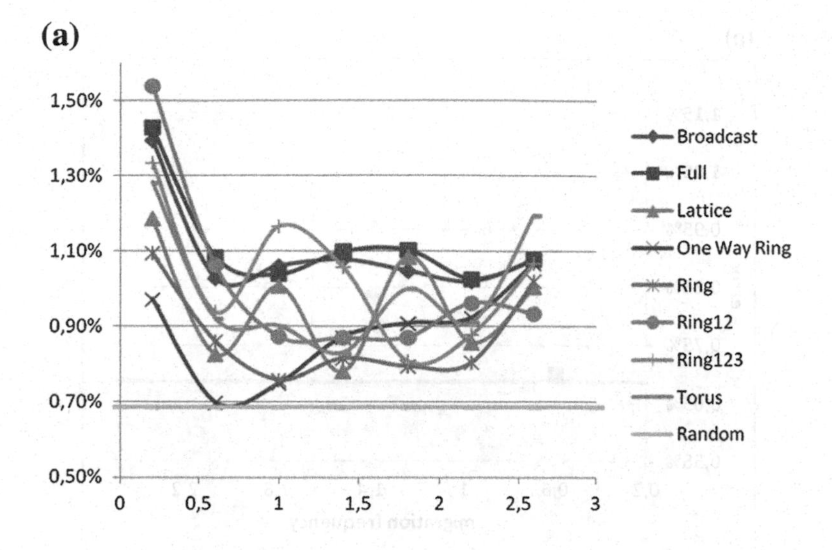

(b)

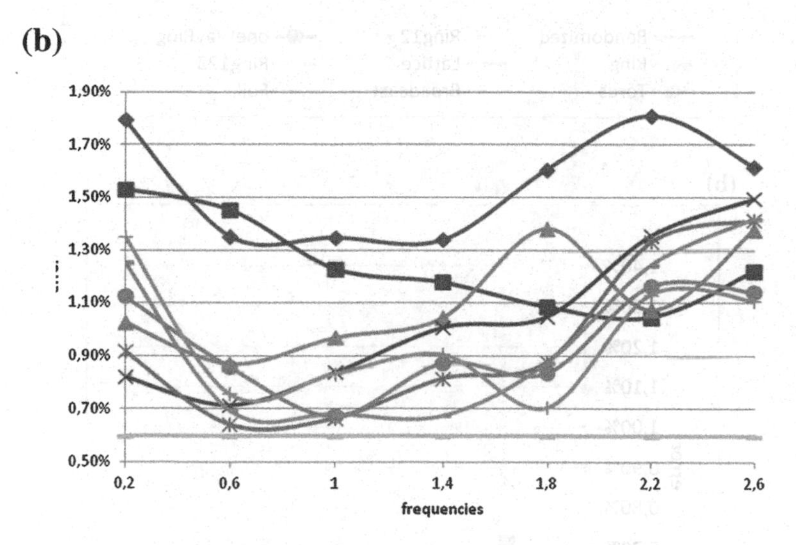

Fig. 8.4 Relative errors averaged over all tasks (**a**) 8 islands (**b**) 16 islands

topology in both cases (for 8 and 16 islands) are better than those for other topologies. The next best topology was *One Way Ring* for 8 islands, and *Ring* for 16 islands, both using frequency 0.6. For both considered numbers of islands the worst results have been obtained for topologies *Broadcast* and *Fully connected*.

Figure 8.6 presents average relative errors for the best topologies: *One Way Ring* and *Ring*, for respective numbers of islands and considered migration frequencies. For comparison, in the same figure additionally the results from *Randomized* topology have been drawn. It may be noticed that the choice of frequency becomes more

(a)

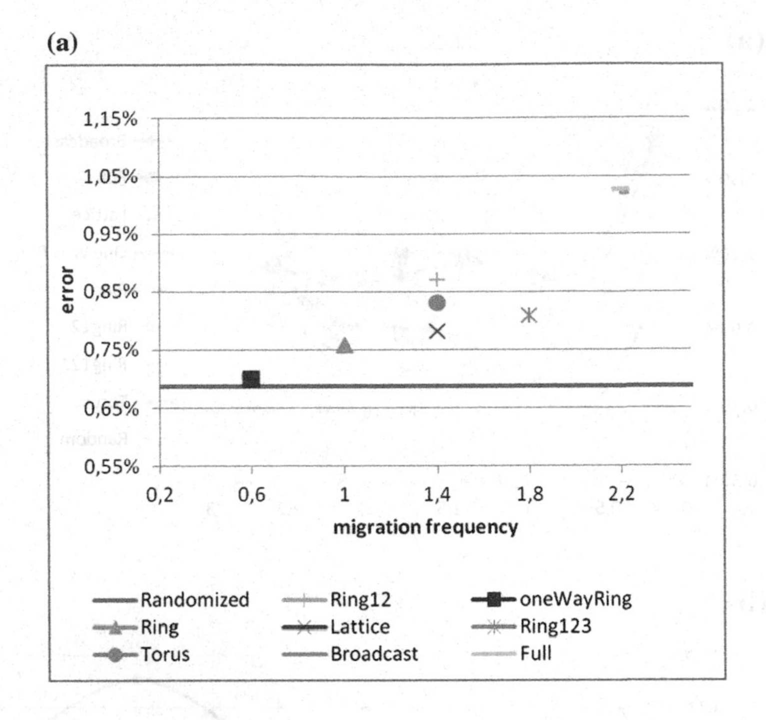

(b)

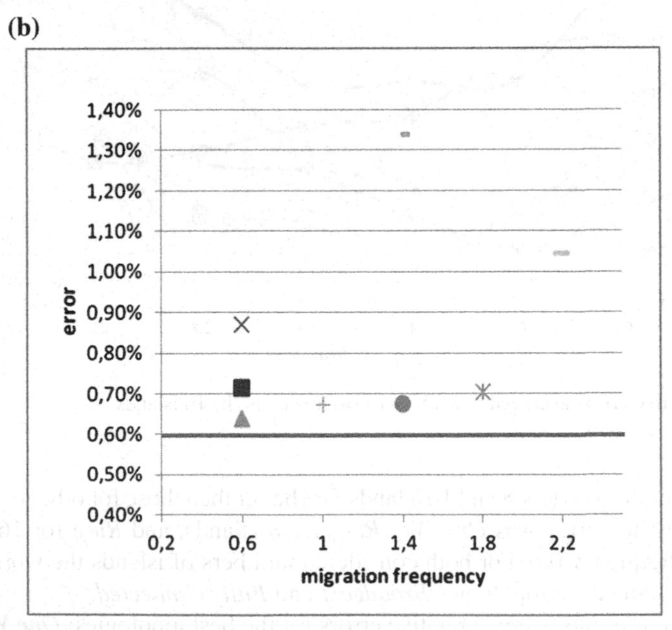

Fig. 8.5 Minimal relative errors for considered migration frequencies and topologies (**a**) 8 islands (**b**) 16 islands

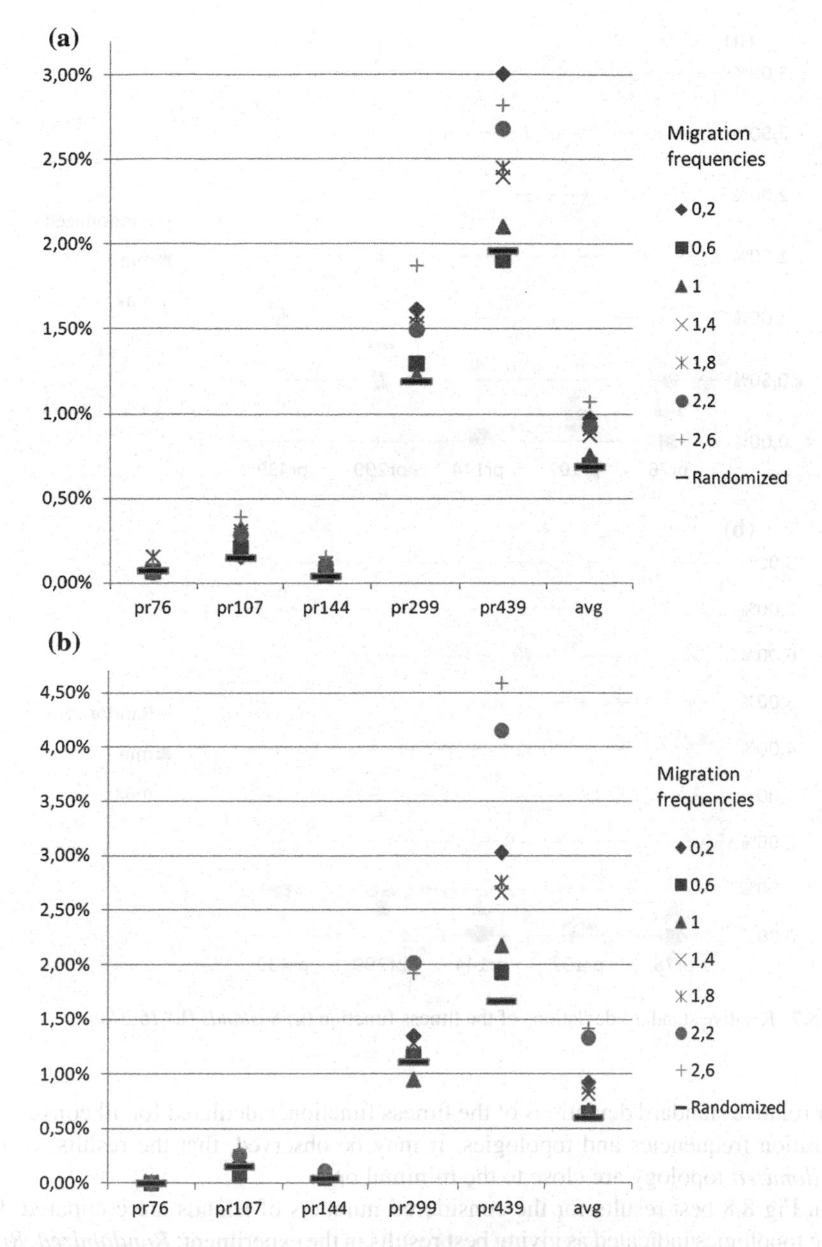

Fig. 8.6 Relative errors for best topologies (**a**) 8 islands, *One Way Ring* and *Randomized* topologies (**b**) 16 islands, *One Way Ring* and *Randomized* topologies

important when the size of the task increases. *Randomized* in comparison with other topologies gives fairly good results independently of the task size.

Figure 8.7 compares *Randomized* topology with other topologies in terms of relative standard deviation. For a given task, minimum and maximum has been taken

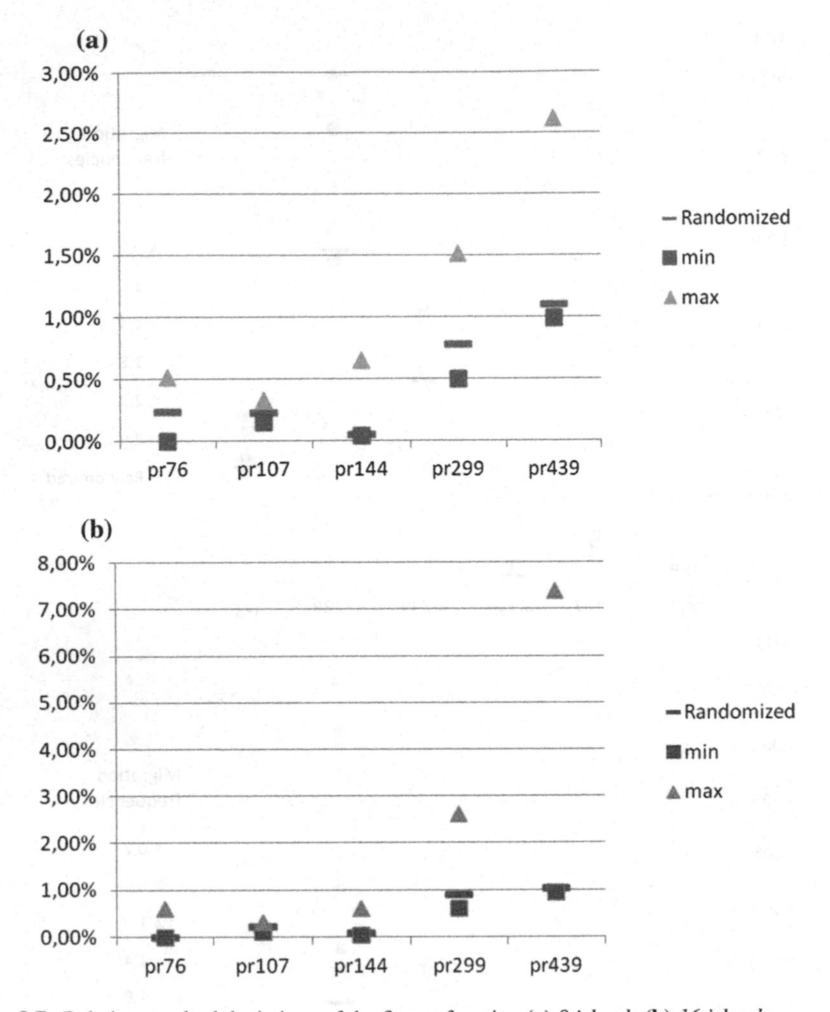

Fig. 8.7 Relative standard deviations of the fitness function (**a**) *8 islands* (**b**) *16 islands*

over relative standard deviations of the fitness function, calculated for all considered migration frequencies and topologies. It may be observed, that the results for the *Randomized* topology are close to the minimal ones.

In Fig. 8.8 best results for the considered numbers of islands are compared, for three topologies indicated as giving best results in the experiment: *Randomized*, *Ring* and *One Way Ring*. The figure shows that not only the choice of topology influences the results, but also the choice of other parameters, like in this case the number of islands. Results obtained with topology *Randomized* always improve when 16 islands is used, as opposed to 8 islands. However it is not a rule for other topologies: for example *One Way Ring* used with the biggest task considered (pr439) gives better results when less islands is used.

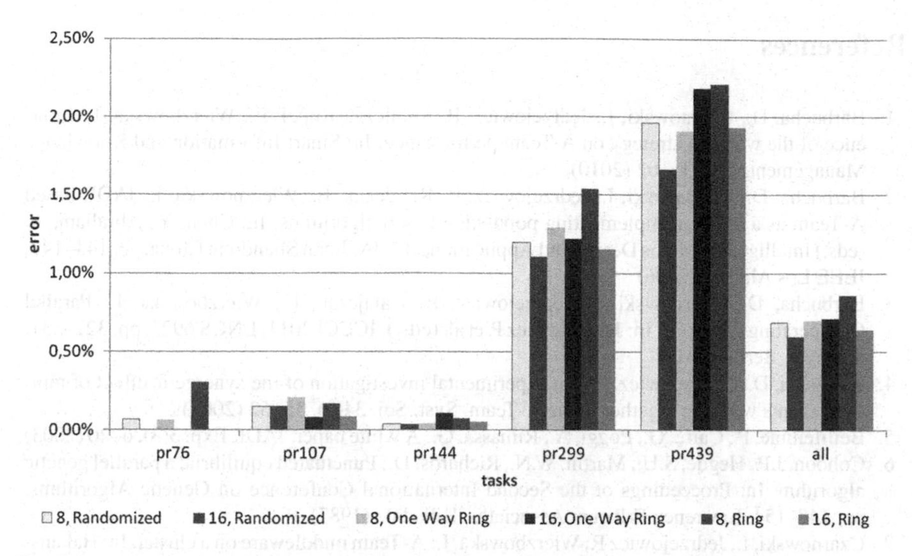

Fig. 8.8 Relative errors for different topologies and numbers of islands, frequency 0.6

It has to be noted, that the computations with 16 islands were obtained in up to 30 % longer calculation time, which may result from reaching limits of capacity of existing resources (in this case significantly increases the number of messages sent within the system). Thus, deciding which topology should be used should go along with checking other parameters of the system. Or, alternatively, the system itself has to be optimised to be able to work with bigger numbers of islands and agents.

8.5 Conclusions

The results of the discussed experiment confirmed that the choice of the migration topology and frequency may influence results obtained by the TA-Teams. Thus, the importance of carefully choosing these two parameters has been confirmed.

The chapter also indicates that the topology called *Randomized* may give significantly better results than other topologies within considered range of migration frequencies. What is more, *Randomized* does not use frequency and thus does not require deciding on the value of this parameter.

The observations are valid only to one problem considered in this chapter, that is EPTSP. Future research may focus on evaluating effects of the choice of migration parameters in solving other problems. Also, the other parameters used in the system may be evaluated—like for example best number of islands or different strategies.

Acknowledgments Calculations have been performed in the Academic Computer Centre TASK in Gdansk. The research has been supported by the Ministry of Science and Higher Education no. N N519 576438 for years 2010-2013.

References

1. Barbucha, D., Czarnowski, I., Jędrzejowicz, P., Ratajczak-Ropel, E., Wierzbowska, I.: Influence of the working strategy on A-Team performance. In: Smart Information and Knowledge Management, pp. 83–102 (2010)
2. Barbucha, D., Czarnowski, I., Jędrzejowicz, P., Ratajczak, E., Wierzbowska, I.: JADE-based A-Team as a tool for implementing population-based algorithms. In: Chen, Y., Abraham, A. (eds.) Intelligent Systems Design and Applications, ISDA, Jinan Shandong China, pp. 144–149. IEEE Los Alamos (2006)
3. Barbucha, D., Czarnowski, I., Jędrzejowicz, P., Ratajczak, E., Wierzbowska, I.: Parallel Co-operating A-Teams. In: Jędrzejowicz P. et al. (eds.): ICCCI 2011, LNCS 6923, pp. 322–331, Springer, Berlin (2011)
4. Barbucha, D., Jędrzejowicz, P.: An experimental investigation of the synergetic effect of multiple agents working together in the A-Team. Syst. Sci. **34**(2), 55–62 (2008)
5. Bellifemine, F., Caire, G., Poggi, A., Rimassa, G.: A white paper. JADE Exp. **3**(3), 6–20 (2003)
6. Cohoon, J.P., Hegde, S.U., Martin, W.N., Richards, D.: Punctuated equilibria: a parallel genetic algorithm. In: Proceedings of the Second International Conference on Genetic Algorithms. pp. 148–154. Lawrence Erlbaum Associates, Hillsdale (1987)
7. Czarnowski, I., Jędrzejowicz P., Wierzbowska, I.: A-Team middleware on a cluster. In: Hakansson A. et al. (eds.) KES - AMSTA 2009, LNAI 5559, pp. 764–772. Springer, Heidelberg (2009)
8. Czarnowski, I., Jędrzejowicz, P.: Agent-based simulated annealing and Tabu search procedures applied to solving the data reduction problem. Int. J. Appl. Math. Comput. Sci. **21**(1), 57–68 (2011)
9. Jędrzejowicz, P., Wierzbowska, I.: JADE-based A-Team environment. In: Alexandrov V.N. et al. (eds.): ICCS 2006, Part III, LNCS 3993, pp. 719–726. Springer, Berlin (2006)
10. Jędrzejowicz, P., Wierzbowska, I.: Experimental investigation of the synergetic effect produced by agents solving together instances of the euclidean planar travelling salesman problem. KES-AMSTA **2**, 160–169 (2010)
11. Jędrzejowicz, P., Wierzbowska, I.: Impact of migration topologies on performance of teams of A-Teams. In: Grana M. et al. (eds) Advances in Knowledge-Based and Intelligent Information and Engineering Systems, pp. 1161–1180. IOS Press (2012)
12. Rucinski, M., Izzo, D., Biscani, F.: On the impact of the migration topology on the Island model. Parallel Comput. **36**(10–11), 555–571 (2010)
13. Talukdar, S.N.: Collaboration rules for autonomous software agents. Decis. Support Syst. **24**, 269–278 (1999)
14. Talukdar, S., Baerentzen, L., Gove, A., de Souza, P.: Asynchronous teams: cooperation schemes for autonomous agents. J. Heuristics **4**(4), 295–332 (1998)
15. Talukdar, S.N., Pyo, S.S., Giras, T.: Asynchronous procedures for parallel processing. IEEE Trans. PAS **PAS-102**(11), 3652–3659 (1983)
16. Talukdar, S.N., de Souza, P., Murthy, S.: Organizations for computer-based agents. Eng. Intell. Syst. **1**(2), 56–69 (1993)
17. Tanese, R.: Distributed genetic algorithms. In: Schaffer, J. (ed.) Proceedings of the Third International Conference on Genetic Algorithms, pp. 434–439. Morgan Kaufmann, San Mateo, USA (1989)
18. Whitley, D., Rana, S., Heckendorn, R.B.: The Island model genetic algorithm: on separability, population size and convergence. J. Comput. Inform. Technol. **7**, 33–47 (1998)
19. Wooldridge, M.: An Introduction to multiagent systems. Wiley, England (2002)

Chapter 9
A Cooperative Agent-Based Multiple Neighborhood Search for the Capacitated Vehicle Routing Problem

Dariusz Barbucha

Abstract The chapter proposes a new hybrid approach for solving Capacitated Vehicle Routing Problem (CVRP), which integrates the cooperative multiple neighborhood search with a multi-agent paradigm. Using the multiple neighborhoods, explored by different heuristics during the search allows one to guide the search and avoid the reaching unsatisfactory results, whenever the search is getting trapped in a local optimum. On the other hand, a multi-agent architecture provides an effective mechanism for solving the problem in parallel and assures cooperation between agents (representing search methods) operating on a sharable population of solutions. Different strategies of exploration of multiple neighborhoods have been considered in the chapter. Some of them search for the best solutions using a family of still deeper neighborhoods, while others use the idea of systematically changing different neighborhoods according to the predefined order (neighborhoods were explored in randomly order or the order of exploration of neighborhoods were based on the neighborhood size). In order to validate the proposed approach a computational experiment has been carried out. It confirmed that using multiple neighborhoods may improve the computational results comparing to the cases, when only one neighborhood is explored during the search.

Keywords Neighborhood Search · Multi-Agent System · Agent-Based Optimization · Cooperation · Vehicle Routing

---

D. Barbucha (✉)
Deptartment of Information Systems, Gdynia Maritime University, Morska 83,
81-225 Gdynia, Poland
e-mail: d.barbucha@wpit.am.gdynia.pl

J. W. Tweedale and L. C. Jain (eds.), *Recent Advances in Knowledge-based Paradigms and Applications*, Advances in Intelligent Systems and Computing 234,
DOI: 10.1007/978-3-319-01649-8_9, © Springer International Publishing Switzerland 2014

9.1 Introduction

Local Search (LS) (or the neighborhood search) methods are one of the most frequently used technique for solving combinatorial optimization problems. These methods iteratively search for the best solution in the solutions space. Starting from an arbitrary candidate solution, LS moves to one of its neighbours, presumably better (in term of the objective function value) than the current candidate solution. If such solution is found, in the next iteration, the process is repeated, with the newly found solution as a starting point of the search. If all neighbours of current solution are inspected and they are inferior to the current solution, then LS stops. The current solution is taken as a local optimum.

This relatively simple scheme of search often produces a solution in a short time, however, the quality of solutions obtained by LS, is often not satisfactory. Hence, the majority of approaches for solving combinatorial optimization problems use local search within a broader frameworks, like, for example, metaheuristics [16]. Among many advantages of such methods, in context of LS, they often allow LS to overcome disadvantage of getting stuck in local optima and enable to continue searching for a better solution by exploration of other promissing regions of search space. Different forms of escaping from local optima suggested in metaheuristics can be: accepting an nonimproved neighbour, as in Simulated Annealing [13], using a modified, penalized goal function in Guided Local Search [34] or, for example, smoothing the search space [20]. In this regard, an interesting idea, based on exploration of the multiple neighborhoods, has been proposed by several authors in approaches, like Variable Neighborhood Search [21], Very Large Scale Neighborhood Search [1, 30], or Adaptive Large Neighborhood Search [29]. This last group of methods has become an inspiraton for the author for research presented in this chapter.

Typical classification of metaheuristics distinguishes: single solution based metaheuristics and population-based ones [31]. The first group of methods, including, for example, Tabu Search [17], Simulated Annealing [13] or Greedy Randomized Adaptive Search [14] procedures, concentrates on improving a single solution (individual). On the other hand, population-based metatheuristics handle a population of individuals that evolves with the help of information exchange procedures. A class of the population based metaheuristics, mostly inspired by biological or social processes, includes Scatter Search methods [18], Evolutionary Algorithms [24], Gene Expression Programming approaches [15], Ant Colony Optimization algorithms [12], and Particle Swarm Optimization algorithms [22].

Last years, an interesting form of hybridization of different methods seems to be *cooperative search*. It consists of a search performed by highly *autonomous programs*, each implementing a particular solution method, working under a *cooperation scheme* which combines these programs into a single consistent problem-solving strategy [7, 11]. Generaly, a set of autonomous programs may include exact methods, like for example branch and bound, but in most cases different approximate algorithms (Local Search, Variable Neighborhood Search, Evolutionary Algorithms and Tabu Search) are engaged in finding the best solution. A cooperation scheme, has to provide the mechanism for effective communication between autonomous programs allowing them to dynamically exchange the important pieces of information which next is used by each of them to support the process of search for a solution.

Nowadays, the natural approach to implementation a cooperative search including various parallel and distributed versions of metaheuristics is design and use multi-agent systems [35]. Such systems, composed of multiple autonomous components (agents) can model complex behavior and can be used to solve real-world problems in a range of industrial and commercial applications. An example of such an approach is a concept of the Asynchronous Team (A-Team) [32], which integrates paradigms of the population-based methods, cooperative problem solving and multi-agent systems.

This chapter aims at proposing a new metaheuristic approach for solving Capacitated Vehicle Routing Problem (CVRP), which integrates the cooperative multiple neighborhoods search with a Multi-agent System (MAS) paradigm. The main motivation is that using the multiple neighborhoods, explored by different heuristics during the search allows one to guide the search and avoid reaching unsatisfactory results. On the other hand, a MAS architecture provides an effective mechanism for solving the problem in parallel and for cooperation between agents (representing search methods) operating on a sharable population of solutions. The chapter extends the author's provious work [4].

The reminder of this chapter is organized as follows. Section 9.2 includes an overview of multiple neighborhood search methods and presents different variants of Vehicle Routing Problem (VRP) with a review of existing multiple neighborhood search approaches for solving them. Section 9.3 focuses on a cooperative agent-based multiple neighborhood search approach for solving CVRP, while in Sect. 9.4 a results of computational experiment which has been carried out in order to validate the proposed approach are presented and discussed. Finally, Sect. 9.5 provides conclusions and suggestions of future work.

9.2 Background

9.2.1 Multiple Neighborhood Search

An idea of exploration of the multiple neighborhoods by one or more, search threads (possibly running in parallel) has been used by several authors in a number of approaches, like Variable Neighborhood Search (VNS) [21], Very Large Scale Neighborhood Search (VLNS) [1, 30], or Adaptive Large Neighborhood Search (ALNS) [29].

VNS is a metaheuristic proposed by Mladenovic and Hansen [25]. Its general idea is a systematic exploration of the set of predefined neighborhoods $N_1, N_2, \ldots, N_{k_{max}}$ during the search process. It changes between neighborhoods, in order to avoid getting trapped in a local optima with poor quality solutions. VNS is based on the fact that the concept of local optimality is conditional on the neighborhood structure used in a local search. Using various neighborhoods in a local search may generate different local optima and the global optimum can be seen as a local optimum for a given neighborhood. Since a local optimum with respect to one neighborhood is not necessarily a local optimum with respect to another neighborhood, changing neighborhoods in the search is a way of diversifying the search.

Apart from the defining the neighborhoods, a major challenge in designing an effective VNS algorithm is to define the order in which the neighborhoods should be searched. The first, natural strategy, is to explore neighborhoods at random. Another one is to order the neighborhoods according to the neighborhood size and/or complexity of exploring them, such that one starts with the simplest neighborhood, and gradually covers the more expensive. Finally, the order of selecting neighborhoods may also be defined by considereing one neighborhood, but with variable depth. Whenever the algorithm reaches a local minimum using one of the neighborhoods, it proceeds with a larger one belonging to the set of neighborhoods.

VLNS [1] is based on the observation that searching a large area results in finding local optima of better quality. Unfortunately, exploration of a large neighborhood is more time consuming, hence various techniques, which restrict the neighborhood search space are used. An interesting derivative subclass of VLNS are Variable-Depth Neighborhood Search (VDNS) methods, which search a parameterized family of neighborhoods $N_1, N_2, \ldots, N_{k_{max}}$ in a heuristic way, gradually extending the size of the neighborhood, each time the search gets trapped in a local minimum (in this context, parameterized family means a family including the same form of neighborhood, but where each neighborhood is an extension of another). Another one is the Large Neighborhood Search (LNS) metaheuristic, proposed by Shaw [30], where the idea of search is based on a gradual improvement of the initial solution by alternately destroying and repairing solutions. Oposite, to the most neighborhood search algorithms, where the neighborhood is defined explicitly, in the LNS metaheuristic the neighborhood is defined implicitly by methods (often heuristics) which are used to destroy and repair an incumbent solution. The neighborhood of a given solution is then defined as the set of solutions that can be reached by first applying the destroy method and then the repair method.

Finally, the ALNS heuristic, proposed by Ropke and Pisinger [29], extends the VLNS heuristic by allowing multiple neighborhoods within the same search and where each destroy/repair method (exploring different neighborhoods) is assigned a weight that controls the frequency of applying the particular method during the search. The weights are adjusted dynamically using recorded performance of the method exploring given neighborhoods so that the heuristic adapts to the instance at hand and to the state of the search.

9.2.2 Vehicle Routing Problems

One of the well known group of computational difficult problems is vehicle routing and storage management [23, 33]. Although problems belonging to this group have a variety of features that differentiate them, all can be modelled as an undirected/directed graph $G = (V, E)$, where $V = \{0, 1, \ldots, n\}$ is the set of nodes and E is a set of edges. Node 0 is a central depot with NV identical vehicles of capacity W and each other node $i \in V \setminus \{0\}$ denotes customer (with its request) with a non-negative demand d_i. Each link (i, j) between two customers denotes the shortest path from customer i to j and is described by the cost c_{ij} of travel from i to j by shortest path $(i, j = 1 \ldots, n)$.

The fundamental goal of VRP is to find vehicle routes which minimize the total cost of travel (or travel distance) and such that each route starts and ends at the depot, each customer is serviced exactly once by a single vehicle, and the total load on any vehicle associated with a given route does not exceed the vehicle capacity. VRP where only the above capacity constraint is imposed is called CVRP [23].

But, in addition to the vehicle capacity constraint, a further limitation can be imposed on the total route duration. In such case t_{ij} is defined to represent the travel time for each edge $(i, j) \in E$, and t_i represents the service time at any vertex i $(i \in V \setminus \{0\})$. It is required that the total duration of any route should not exceed a preset bound T.

An extension of the VRP includes Vehicle Routing Problem with Time Windows (VRPTW) [8, 9], where for each delivery location a time window $[e_i, l_i]$ is defined, within which the deliveries (or visits) must be made. Here, e_i defines earliest allowed arrival time (opening time), and l_i is the latest allowed arrival time (closing time). If no arrivals are allowed outside of the given parameters, the time windows is said to be hard. On the other hand, the soft time window is defined, when delivery is allowed after l_i. In addition to minimize the total transportation costs, one of the other goal considered in this variant is also a minmization of the total vehicles used.

In turn, Pickup and Delivery Vehicle Routing Problem (PDVRP) [26, 27] assumes that a number of locations is defined as pickup, and another as delivery locations. All goods need to be moved from certain pickup to other delivery locations. By adding time windows constraints to the PDVRP, one can define Pickup and Delivery Vehicle Routing Problem with Time Windows (PDVRPTW). And, taking into account, for example, characteristics of the fleet of vehicles, a Heterogeneous Fleet Vehicle Routing Problem (HVRP) [2] can be distinguished. A review of the different variants of VRP and state-of-the-art methods for solving them, can be found, for example, in [19, 23].

Because of the NP-hardness of VRP, solution methods designed for solving different instances of the problem have mainly heuristic nature. A group of multiple neighborhood search methods (including LNS and ALNS) also successfully find an application in solving them. There are many examples of applications of multiple neighborhood search (and its different special cases) to VRP variants, and what is important, many of them have been successful and have provided state-of-the-art results at the time of publication [28].

Shaw [30], after introducing an idea of the LNS heuristic, he illustrated its behaviour and performance for the vehicle routing problem. The proposed LNS for VRP explores a large neighborhood of the current solution by selecting a number of customer visits to remove from the routing plan, and next re-inserting these visits using a constraint-based tree search. His approach also maintains diversity during search by dynamically altering the number of visits to be removed, and by using a randomised choice method for selecting visits to remove.

The other representative methods have been proposed by Bent and van Hentenryck in their research for VRPTW [5] and PDVRPTW [6]. In both, they propose to solve the problem in a two-stage approach. In the first stage the number of routes is minimized by a simulated annealing algorithm that uses traditional, small neighborhoods. In the second stage the total route lengths are minimized with an LNS heuristic. The size of the neighborhood is gradually increased, starting out by only removing one customer and by steadily increasing the number of customers to remove as the search progresses. At regular intervals, the number of customers to remove is reset to one and the neighborhood size increase starts over.

Another representative example is an approach of Ropke and Pisinger [29] which introduced the ALNS for the PDVRPTW. Compared to the LNS heuristic developed for the VRPTW and PDVRPTW by Shaw [30] and Bent and van Hentenryck [5, 6], the heuristic of Ropke and Pisinger is different in several ways. First, they use several removal and insertion heuristics during the same search while the earlier LNS heuristics only used one method for removal and one method for insertions. Moreover, during the search, their approach chooses between these heuristics using statistics gathered during the search. Next, simple and fast heuristics are used for the insertion of requests as opposed to the more complicated branch and bound methods proposed by Shaw [30] and Bent and van Hentenryck [5, 6]. Finally, the search is embedded in a simulated annealing metaheuristic where the earlier LNS heuristics used a simple descent approach.

The approach proposed in this chapter focuses on applying a cooperative multiple neighborhood search working within a multi-agent framework for the CVRP.

9.3 A Cooperative Agent-Based Multiple Neighborhood Search for the Capacitated Vehicle Routing Problem

Technically, the proposed approach for solving CVRP is based on a middleware supporting a construction of the dedicated A-Team architectures [32] used for solving a variety of computationally hard optimization problems [3]. Its main components are presented in Fig. 9.1.

The approach produces solutions to combinatorial optimization problems using a *common, sharable memory*, which store a population of individuals (solutions) and a *set of agents*, each representing an improvement algorithm, which operate on individuals during the process of search. Moreover, a specialy designed program, called solution manager, acts as an intermediary between common memory and multiple neighborhood search programs. It maintains the common memory and its role is to read a particular individual from the memory and to send it periodically to search procedures, which have already announced their readiness to act, and to

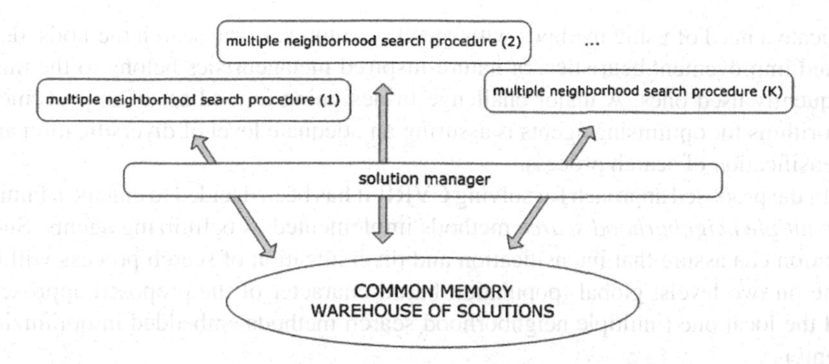

Fig. 9.1 Main components of the architecture of the proposed approach

update the memory by storing in it a possibly improved solution obtained from the search program.

At first, the process of search starts from a step of generation of an initial population of solutions. Next, the main steps of the proposed approach, repeated until a stopping criterion is met, include:

1. Selecting a particular individual (solution) from the common memory and sending it to the autonomous, independently acting optimising agents, representing multiple neighborhood search procedures,
2. Improvement of solutions by these agents, and
3. Storing back the potentially improved solution returned by an optimising agent in the common memory.

Throughout the whole process of solving an instance of the problem, the memory succesively evolves from the initially generated pool of solutions through intermediate trial solutions obtained during the search process up to the stage when the stopping criterion is met, and the best solution stored in the population is taken as the final solution of the given problem instance.

The existence, within A-Team, of shared memory, a mechanism of management of population of solutions and a set of autonomous agents, provide a basis for cooperation between agents. Solutions obtained by one agent are shared, through the central memory mechanism, with other agents, which can exploit these solutions in order to guide the search through new promising region of the search space, thus increasing chances for reaching the global optimum. It is expected that such a collective of agents can produce better solutions than individual members of such collective, thus, achieving a synergetic effect.

Another important advantage of applications of the A-Team concept for solving particular optimization problem, stems from a problem-solving method encapsulated in each optimization agent. Although the whole approach belongs to a group of the population-based methods, each optimizing agent is, in fact, an implementation of a single-solution method. The general assumptions about A-Team do not indicate a need of using methods with specific features. Local search methods, dedicated improvement heuristics, or nature-inspired metaheurisics belong to the most frequently used ones. A major challenge in designing a good set of improvement algorithms for optimising agents is assuring an adequate level of diversification and intensification of search process.

In the proposed approach for solving CVRP, it has been decided to engage a family of *multiple neighborhood search* methods implemented as optimizing agents. Such solution can assure that intensification and diversification of search process will be done on two levels: global (population-based character of the proposed approach) and the local one (multiple neighborhood search methods embedded in optimizing agents).

Similar to other population-based methods, in order to use the proposed approach for solving particular problem, definitions of a few elements are required. These include: representation of individual, method of creating an initial population, fitness, method of managing a population of individuals, and a set of optimizing agents, representing multiple neighborhood search methods.

In the proposed approach, each solution of the problem is represented as individual in population and has a permutation *form* of n numbers (representing customers) form with additional '**0**' delimiters denoting division of permutation into the routes. A part of individual between '**0**' delimiters reflects the order in which customers are visited by one vehicle within selected route. An individual would be represented, for example, as: [**0** 3 7 9 **0** 1 5 6 2 **0** 4 8 **0**].

Population of individuals is created randomly, which means that, first, permutation of n numbers, representing each individual is created randomly, and next, '0' delimiters are inserted in places which divide the permutation into the separeted parts (routes) assigned to each vehicle. The places for delimiters insertion is calculated in such a way that total capacity of vehicle assigned to the current route and the maximal route length are not exceeded. The process of creating the whole initial population is repeated until PS (population size) individuals have been generated.

Each individual from the population is evaluated using the *fitness* function, which value is calculated as a sum of the costs related to each permutation part (vehicle's route).

Three sets of *optimizing agents* have been proposed, which explore the search space using different strategies of switching between neighborhoods:

- OA_{VDNS}: A set of optimizing agents based on the variable-depth neighborhood search with a predefined family of still deeper neighborhoods,
- OA_{VNS}: A set of optimizing agents representing the variable neighborhood search where the order of neighborhood exploration is defined according to the neighborhood size,
- OA_{RAND}: A set of optimizing agents representing the variable neighborhood search where the neighborhoods are explored in a random order.

Having a finite set of pre-selected neighborhood structures $N = \{N_1, N_2, \dots, N_{k_{max}}\}$, all the above kinds of agents initially begin their search from the first neighborhood ($k = 1$). If the local search leads to a new best solution, then k is reset to 1,

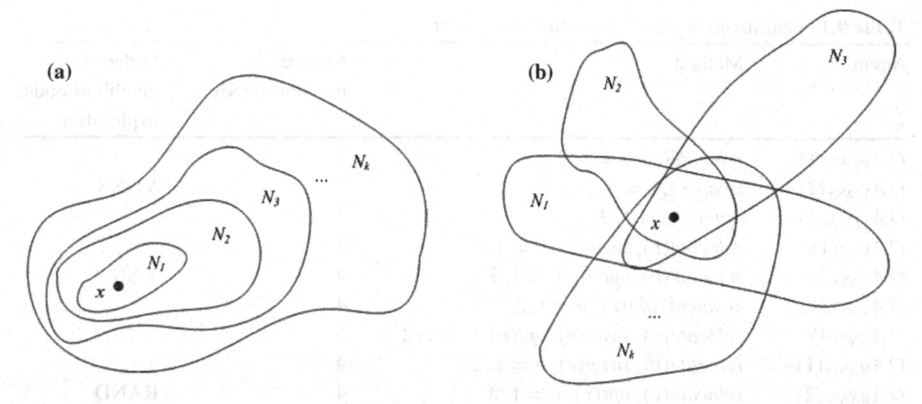

Fig. 9.2 Description of a set of neighborhoods used by optimizing agents. **(a)** OA_{VDNS} and **(b)** OA_{VNS} and OA_{RAND}

otherwise k is increased by one (for OA_{VDNS} and OA_{VNS}), or k is chosen randomly (for OA_{RAND}). Each agent stops its search where no improvement is observed for all neighborhoods. A set of neighborhoods and its structure used by the above agents are shown in Fig. 9.2.

Finally, *management of population of individuals* stored in the common memory is supervised by solution manager and is performed in a loop, where a cycle of the selecting-improving-replacing is performed several times until a predefined time (from last improvement of the best solution) has elapsed. The step of *selecting* an individual from the common memory and next sending it to the optmising agents is implemented as a selection of a random solution. After improvement phase, if the solution currently received from optimization agent has been improved, it is *accepted* and *replaces* the worst solution from current population. Additionally, if last consecutive five solutions received from the optimization agents did not improve existing solutions in population, the worst solution is removed from the population and a newly generated one is added to the pool of individuals.

9.4 Computational Experiment

Computational experiment has been carried out to validate effectiveness of the proposed approach while solving instances of the CVRP. The experiment aimed at answering the question: To what extent (if any) different forms of multiple neighborhoods search incorporated into agent-based framework influence computation results? The quality of the results obtained by the proposed approach has been measured as the MRE from the optimal (or the best known) solution, reported by Laporte et al. [23].

Table 9.1 Optimising agents used in the experiment

Agent	Methods	Number of neighborhoods	Order of neighborhoods exploration
$OA_{VDNS}(1)$	$relocate(v)$, $v = 1, 2$	2	
$OA_{VDNS}(2)$	$swap(v)$, $v = 1, 2$	2	VDNS
$OA_{VDNS}(3)$	$opt(v)$, $v = 1, 2$	2	
$OA_{VNS}(1)$	$relocate(v)$, $swap(v)$, $v = 1, 2$	4	
$OA_{VNS}(2)$	$relocate(v)$, $opt(v)$, $v = 1, 2$	4	VNS
$OA_{VNS}(3)$	$swap(v)$, $opt(v)$, $v = 1, 2$	4	
$OA_{VNS}(4)$	$relocate(v)$, $swap(v)$, $opt(v)$, $v = 1, 2$	6	
$OA_{RAND}(1)$	$relocate(v)$, $swap(v)$, $v = 1, 2$	4	
$OA_{RAND}(2)$	$relocate(v)$, $opt(v)$, $v = 1, 2$	4	RAND
$OA_{RAND}(3)$	$swap(v)$, $opt(v)$, $v = 1, 2$	4	
$OA_{RAND}(4)$	$relocate(v)$, $swap(v)$, $opt(v)$, $v = 1, 2$	6	

Eleven optimizing agents, belonging to three previously defined sets, and representing different multiple neighborhood search methods have been proposed and presented in Table 9.1.

All agents include combination of three groups of methods, each exploring two neighborhoods:

- a group of two methods $relocate(v)$, $v = 1, 2$, which explore two neighborhoods using moves which relocate 1 or 2 customers, respectively, from their original positions to another ones.
- a group of two methods $swap(v)$, $v = 1, 2$, which explore two neighborhoods using moves which choose 1 or 2 pairs of customers, respectively, and swap customers within each pair.
- a group of two methods $opt(v)$, $v = 1, 2$, which explore two neighborhoods using moves which remove 2 or 3 edges, respectively (each edge includes two succesive customers) forming 2 or 3, respectively, disconnected segments and next reconnect these segments in all possible ways.

Combining the above agents, different teams are possible to construct. The following cases, where a performance of selected teams of agents are compared, have been considered in the experiment:

Case #1: $OA_{VNS}(1)$ versus $OA_{VDNS}(1) + OA_{VDNS}(2)$
Case #2: $OA_{VNS}(2)$ versus $OA_{VDNS}(1) + OA_{VDNS}(3)$
Case #3: $OA_{VNS}(3)$ versus $OA_{VDNS}(2) + OA_{VDNS}(3)$
Case #4: $OA_{VNS}(4)$ versus $OA_{VDNS}(1) + OA_{VDNS}(2) + OA_{VDNS}(3)$
Case #5: $OA_{RAND}(1)$ versus $OA_{VDNS}(1) + OA_{VDNS}(2)$
Case #6: $OA_{RAND}(2)$ versus $OA_{VDNS}(1) + OA_{VDNS}(3)$
Case #7: $OA_{RAND}(3)$ versus $OA_{VDNS}(2) + OA_{VDNS}(3)$
Case #8: $OA_{RAND}(4)$ versus $OA_{VDNS}(1) + OA_{VDNS}(2) + OA_{VDNS}(3)$

In all cases, teams consisting of 2 or 3 optimising agents representing VDNS methods are compared with one optimizing agent, including the same methods, but integrated within one VNS method. Whereas the order of neighborhoods exploration in VNS agents is random in cases #5 to #8, it is defined according to the neighborhood size in cases #1 to #4.

Seven CVRP instances of Christofides et al. [10] used in the experiment (*vrpnc1-vrpnc5*, *vrpnc11-vrpnc12*) contain 50 to 199 customers. For each case #1 to #8, each instance was solved 10 times, in total giving 560 (7*8*10) test problems. Population size was set to 30 individuals. The system stops after 180 seconds of running without an improvement.

All computations have been carried out on the computer cluster called HOLK of the Tricity Academic Computer Network. This contains 265 Intel Itanium 2 nodes (each has 2 Dual Core processors clocked at 1.4 GHz with 12 MB L3 cache) with 2.3 TB total system memory and 4 TB disk storage. Mellanox InfiniBand interconnect with 10 Gb/s bandwidth assures connection of all nodes to the network.

The performance (MRE in %) of the proposed agent-based approach for all cases and all considered instances is presented in Table 9.2 and Fig. 9.3.

Analysis of the results presented in Table 9.2 and Fig. 9.3 provides the first observation that introducing exploration of multiple neighborgoods by local search heuristics increases the ability of producing competitive results by the proposed approach. Even a single agent, which includes only one of the proposed group of methods (*relocate()*, *swap()* or *opt()*) performs well, hovewer its behavior depends on the instance of the problem. Whereas the smallest MRE is observed for instances with a small number of customers, it increases when the size of the problem grows.

Next, integration of two or three of above groups of methods within a single agent creates opportunities for further improving the solution. Such agent can explore a number of neighborhoods equal to the sum of neighborhoods explored be separate methods. For almost all instances, the experiment has confirmed that such approach outperforms the case where each group of methods is run by separate agents, but withing a team.

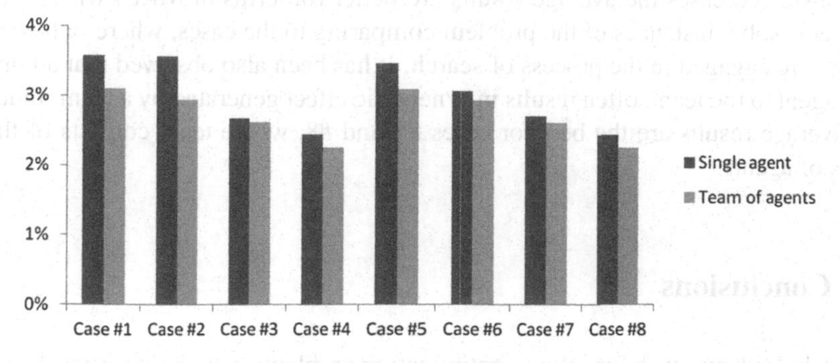

Fig. 9.3 MRE (in %) from the best known solution calculated separately for cases where a single agent or a team of optimizing agents are engaged in process of searching

Table 9.2 MRE (in %) from the best known solution calculated separately for cases where a single agent or a team of optimizing agents are engaged in process of searching

Case	Instance							Average
	vrpnc1 (50)	*vrpnc2* (75)	*vrpnc3* (100)	*vrpnc4* (150)	*vrpnc5* (199)	*vrpnc11* (100)	*vrpnc12* (120)	
Case #1: $OA_{VNS}(1)$ versus $OA_{VDNS}(1)$+$OA_{VDNS}(2)$								
	0.26	1.77	3.01	4.03	6.92	7.81	1.21	3.57
	0.11	1.03	2.78	4.44	5.21	6.91	1.21	3.10
Case #2: $OA_{VNS}(2)$ versus $OA_{VDNS}(1)$+$OA_{VDNS}(3)$								
	0.18	1.28	2.95	4.12	7.79	6.81	1.81	3.56
	0.06	1.61	2.81	3.81	5.12	5.73	1.39	2.93
Case #3: $OA_{VNS}(3)$ versus $OA_{VDNS}(2)$+$OA_{VDNS}(3)$								
	0.19	1.65	2.32	2.57	5.09	4.30	2.59	2.67
	0.06	1.88	1.99	2.90	4.71	5.24	1.57	2.62
Case #4: $OA_{VNS}(4)$ versus $OA_{VDNS}(1)$+$OA_{VDNS}(2)$+$OA_{VDNS}(3)$								
	0.23	1.28	2.27	3.76	4.49	4.23	0.87	2.45
	0.01	0.94	1.91	3.50	4.10	4.11	1.31	2.27
Case #5: $OA_{RAND}(1)$ versus $OA_{VDNS}(1)$+$OA_{VDNS}(2)$								
	0.01	2.21	2.56	5.18	6.78	5.48	2.04	3.47
	0.11	1.03	2.78	4.44	5.91	6.91	1.21	3.10
Case #6: $OA_{RAND}(2)$ versus $OA_{VDNS}(1)$+$OA_{VDNS}(3)$								
	0.19	1.55	3.02	4.28	5.23	5.54	1.65	3.07
	0.06	1.61	2.81	3.81	5.12	5.73	1.39	2.93
Case #7: $OA_{RAND}(3)$ versus $OA_{VDNS}(2)$+$OA_{VDNS}(3)$								
	0.19	1.62	2.54	3.91	3.51	5.90	1.32	2.71
	0.06	1.88	1.99	2.90	4.71	5.24	1.57	2.62
Case #8: $OA_{RAND}(4)$ versus $OA_{VDNS}(1)$+$OA_{VDNS}(2)$+$OA_{VDNS}(3)$								
	0.19	1.94	2.07	3.82	3.08	4.89	1.16	2.45
	0.00	1.73	1.91	3.50	4.10	4.11	0.67	2.27

Also, possibility of running different agents within a team, additionally, reinforces the strength of the proposed approach based on multiple neighborhood search. For all considered cases the average results are better (in terms of MRE) where teams of agents solve instances of the problem comparing to the cases, where only single agents are engaged in the process of search. It has been also observed that adding a new agent to the team, often results in synergetic effect generated by a team. Indeed, the average results are the best for cases #4 and #8, where team consists of three types of agents.

9.5 Conclusions

A new hybrid approach for solving optimization problems, which integrates the multiple neighborhood search with a MAS paradigm has been proposed in the chapter. Whereas the main idea of using the multiple neighborhoods, explored by different heuristics assures diversification of the search, embedding the search in a mult-agent environment has allowed for parallel computations, where agents (representing search methods) operating on a sharable population of solutions, have cooperated during the search.

Different strategies of exploration of multiple neighborhoods have been considered. Although the main goal of them are the same: to avoid getting stuck in poor quality solutions, a major difference between them was the philosophy underlying the proposed approaches. Some of them had the explicit goal of searching a family of still deeper neighborhoods, while others were derived from the idea of systematically changing different neighborhoods during the search, where neighborhoods were explored in randomly order and the order of exploration of neighborhoods were defined according to the neighborhood size.

A computational experiment, which has been carried out using instances of the CVRP, has confirmed that using multiple neighborhoods may improve the computational results comparing to the cases, when only one neighborhood is explored during the search. The overall results can be considered satisfactory and competitive to other results for CVRP, especially, where teams of cooperating agents are considered.

The future research will aim at the implementation of other moves between solutions used in local search inside a single neighborhood type and other rules of switching between neighborhoods. Another interesting direction of research is a using of the proposed agent-based multiple neighborhood search for solving other optimization problems. It will require an implementation of local search methods dedicated for particular problem.

Acknowledgments The research has been supported by the Polish National Science Centre grant no. 2011/01/B/ST6/06986 (2011-2013). Calculations have been performed in the Academic Computer Centre TASK in Gdansk, Poland.

References

1. Ahuja, R.K., Ergun, O., Orlin, J.B., Punnen, A.P.: A survey of very large-scale neighborhood search techniques. Discrete Appl. Math. **123**, 75–102 (2002)
2. Baldacci, R., Battarra, M., Vigo, D.: Routing a heterogeneous fleet of vehicles. In: Golden, B., Raghavan, S., Wasil, E. (eds.) The Vehicle Routing Problem: Latest Advances and New Challenges, pp. 3–28. Springer, Berlin-Heidelberg (2008)
3. Barbucha, D., Czarnowski, I., Jędrzejowicz, P., Ratajczak-Ropel, E., Wierzbowska, I.: JABAT Middleware as a Tool for Solving Optimization Problems. Transactions on Computational Collective Intelligence II. LNCS, vol. 6450, pp. 181–195. Springer, Berlin Heidelberg (2010)
4. Barbucha, D.: An agent-based implementation of the multiple neighborhood search for the capacitated vehicle routing problem. In: Grana, M., Toro, C., Posada, J., Howlett, R.J., and Jain, L.C. (eds.) Advances in Knowledge-Based and Intelligent Information and Engineering Systems. Frontiers in Artificial Intelligence and Applications, vol. 243, pp. 1191–1200. IOS Press (2012)
5. Bent, R., Van Hentenryck, P.: A two-stage hybrid local search for the vehicle routing problem with time windows. Transp. Sci. **38**(4), 515–530 (2004)

6. Bent, R., Van Hentenryck, P.: A two-stage hybrid algorithm for pickup and delivery vehicle routing problem with time windows. Comput. Oper. Res. **33**(4), 875–893 (2006)
7. Blum, C., Roli, A.: Metaheuristics in combinatorial optimization: overview and conceptual comparison. ACM Comput. Surv. **35**(3), 268–308 (2003)
8. Braysy, O., Gendreau, M.: Vehicle routing problem with time windows, Part I: Route construction and local search algorithms. Transp. Sci. **39**, 104–118 (2005)
9. Braysy, O., Gendreau, M.: Vehicle routing problem with time windows, Part II: Metaheuristics. Transp. Sci. **39**, 119–139 (2005)
10. Christofides, N., Mingozzi, A., Toth, P., Sandi, C. (eds.): Combinatorial Optimization. John Wiley, Chichester (1979)
11. Crainic, T.G., Toulouse, M.: Explicit and emergent cooperation schemes for search algorithms. In: Maniezzo, V., Battiti, R., Watson, J.P. (eds.) Learning and Intelligent Optimization (LION II) Conference, LNCS 5313, pp. 95–109. Springer, Berlin (2008)
12. Dorigo, M., Stutzle, T.: Ant Colony Optimization. MIT Press, Cambridge (2004)
13. Eglese, R.W.: Simulated annealing: a tool for operational research. Eur. J. Oper. Res. **46**, 271–281 (1990)
14. Feo, T.A., Resende, M.G.C.: Greedy randomized adaptive search procedures. J. Global Optim. **6**, 109–133 (1995)
15. Ferreira, C.: Gene Expression Programming: Mathematical Modeling by an Artificial Intelligence. Springer, Heidelberg (2006)
16. Gendreau, M., Potvin, J.-Y. (eds.): Handbook of metaheuristics. International Series in Operations Research & Management Science, vol. 146. Springer, New York (2010)
17. Glover, F., Laguna, M.: Tabu Search. Kluwer, Boston (1997)
18. Glover, F., Laguna, M., Marti, R.: Fundamentals of scatter search and path relinking. Control Cybern. **39**, 653–684 (2000)
19. Golden, B.L., Raghavan, S., Wasil, E.A. (eds.): The Vehicle Routing Problem: Latest Advances and New Challenges. Operations Research Computer Science Interfaces Series, vol. 43, Springer, New York (2008)
20. Gu, J., Huang, X.: Efficient local search with search space smoothing: a case study of the traveling salesman problem. IEEE Trans. Syst. Man Cybern. **24**(5), 728–735 (1994)
21. Hansen, P., Mladenovic, N., Brimberg, J., Moreno Perez, J.A.: Variable neighborhood search. In: Gendreau, M., Potvin, J-Y. (eds.) Handbook of Metaheuristics, International Series in Operations Research & Management Science, vol. 146, pp. 61–86. Springer, New York (2010)
22. Kennedy, J., Eberhart, R.: Particle swarm optimization. Proceedings of IEEE International Conference on Neural Networks IV, pp. 1942–1948 (1995)
23. Laporte, G., Gendreau, M., Potvin, J., Semet, F.: Classical and modern heuristics for the vehicle routing problem. Int. Trans. Oper. Res. **7**, 285–300 (2000)
24. Michalewicz, Z.: Genetic Algorithms + Data Structures = Evolution Programs. Springer, New York (1994)
25. Mladenovic, N., Hansen, P.: Variable neighborhood search. Comput. Oper. Res. **24**, 1097–1100 (1997)
26. Parragh, S.N., Doerner, K.F., Hartl, R.F.: A survey on pickup and delivery problems, Part I: Transportation between customers and depot. J. fur Betriebswirtschaft **58**, 21–51 (2008)
27. Parragh, S.N., Doerner, K.F., Hartl, R.F.: A survey on pickup and delivery problems, Part II: Transportation between pickup and delivery locations. J. fur Betriebswirtschaft **58**, 81–117 (2008)
28. Pisinger, D., Ropke, S.: Large neighborhood search. In: Gendreau, M., Potvin, J-Y. (eds.) Handbook of Metaheuristics, International Series in Operations Research & Management Science, vol. 146, pp. 399–419. Springer, New York (2010)
29. Ropke, S., Pisinger, D.: An adaptive large neighborhood search heuristic for the pickup and delivery problem with time windows. Transp. Sci. **40**(4), 455–472 (2006)
30. Shaw, P.: Using constraint programming and local search methods to solve vehicle routing problems. In: Proceedings of Fourth International Conference on Principles and Practice of Constraint Programming CP-98. LNCS, vol. 1520, pp. 417–431 (1998)

31. Talbi, E.G.: Metaheuristics: From Design to Implementation. John Wiley and Sons, Inc., New Jersey (2009)
32. Talukdar, S., Baeretzen, L., Gove, A., de Souza, P.: Asynchronous teams: Cooperation schemes for autonomous agents. J. Heuristics **4**, 295–321 (1998)
33. Toth, P., Vigo, D. (eds.): The Vehicle Routing Problem. Monographs on Discrete Mathematics and Applications. SIAM, Philadelpia (2002)
34. Voudouris, C., Tsang, E.: Guided local search and its application to the traveling salesman problem. Eur. J. Oper. Res. **113**, 469–499 (1999)
35. Wooldridge, M.: An Introduction to MultiAgent Systems. John Wiley and Sons, Chichester (2009)

Chapter 10
Building the "Automatic Body Condition Assessment System" (ABiCA), an Automatic Body Condition Scoring System using Active Shape Models and Machine Learning

Rafael Tedín, José A. Becerra and Richard J. Duro

Abstract A step by step reconstruction of the process that has been followed for building the Automatic Body Condition Assessment (ABiCA) system is presented. ABiCA is an automatic body condition scoring system for dairy cattle using images taken using hand-held cameras. The problem is decomposed into two sub-problems that are solved separately. Firstly, the shape of a cow is found and then the body condition score is estimated using this shape. The solutions to those problems are then combined to build the system. The shape of a cow is found using Active Shape Model (ASMs) tuned with an evolutionary algorithm. The shape feeds then a symbolic regression function evolved by means of genetic programming to finally estimate the body condition score of the cow. The error of the ABiCA system is reasonable, given the uncertainties of the expert's scores. There is nevertheless room for improvement since it has been observed that some images might be too difficult for the system. Methods on how to automatically discard those images are being investigated.

Keywords Active Shape Model · Body Condition Score · Evolutionary Algorithm · Genetic Programming · Machine Learning

---

R. Tedín (✉) · J. A. Becerra · R. J. Duro
Integrated Group for Engineering Research, University of A Coruña, Mendizabal s/n, 15403 Ferrol, (A Coruña), Spain
e-mail: rafael.tedin@udc.es

J. A. Becerra
e-mail: ronin@udc.es

R. J. Duro
e-mail: richard@udc.es

J. W. Tweedale and L. C. Jain (eds.), *Recent Advances in Knowledge-based Paradigms and Applications*, Advances in Intelligent Systems and Computing 234,
DOI: 10.1007/978-3-319-01649-8_10, © Springer International Publishing Switzerland 2014

10.1 Introduction

The milk sector is very relevant in many countries due to its weight within the food industry. To be competitive, this sector has incorporated in recent years all kinds of improvements to maximise the productivity of the farms and to improve the milk production of each cow. Simultaneously, most producers are striving to minimise other costs such as fuel for machinery and purchase of fodder, among others.

To achieve these improvements it is important to be able to collect detailed information about the health of the cows over a long period of time in each farm. Since every individual follows a production–reproduction[1] cycle that is dependent on their overall health, the information gathered permits making estimations of the production of milk within this cycle.

One of these health signs is the Body Condition Score (BCS), which approximately indicates the energy reserves of an animal using a scoring system. Currently, there are different international body scoring systems [1]. In this work the scoring system employed by the experts of the farms that have been visited is used. In that system the BCS is a number in a scale that spans from 1 to 5 (in 0.25 increments).

Ideally, the BCS should be re-evaluated on a regular basis on the farms. A single observation of the BCS is indeed meaningless and the evolution of this score is the relevant information that is needed. Since the goal is to assess the overall state of the farm, so that proper planning and management can be carried out, the progression of a sufficient number of animals has to be tracked. This in turn can be very time consuming, so that an automatic method for scoring is needed.

Currently, the authors are not aware of any procedure that estimates the BCS from images using automated tools. All previous research reviewed need at least a manual segmentation step for the contour of the cow or other structures. Additionally, some methods require costly hardware or fixed set-ups. This is not always possible, since the farms are generally too small for the space requirements of the set-ups and the equipment can be expensive.

The objective of this work is to build ABiCA, an automatic body condition scoring system that can estimate the BCS of cows from images in a reliable and objective manner, in contrast to the subjective estimations of the human experts. In this work the current progress towards achieving this objective is shown.

The text is organised as follows. Section 10.2 presents a more detailed background on automatic body condition scoring for dairy cattle. Then the steps for building the ABiCA system are shown. Section 10.3 is devoted to how an acceptable error for the system has been established and to the feasibility of an automatic body condition scoring system using images. The problem of body condition scoring is decomposed into three sub-problems. For the sake of a better understanding of these problems, the background knowledge for solving each of them is given in the section were the problem is presented. The first sub-problem is finding the representation of the cow (its back shape) given an image of the rear-end of the cow. Active Shape Model

[1] In order to deliver milk (production), every dairy cow has to get pregnant from time to time (reproduction).

(ASMs) [2] are used for this task. The procedure is explained in Sect. 10.4. The second sub-problem is to train a model that, given a representation of a cow, estimates the BCS from it. How this sub-problem can be solved using Machine Learning (ML) techniques is described in Sect. 10.5. The last issue is combining the solutions of the other previous problems to build the complete system. This issue and the obtained results are addressed in Sect. 10.6. Finally, Sect. 10.7 summarises the conclusions of this contribution.

10.2 Background

The BCS is known by field experts as an important value for dairy farms [1]. Several issues related to the health of the animal can be determined by its observation. Examples of such problems are the maximum amount of milk that a dairy cow should produce, the correct evolution of gestation or whether the food supply is adequate, among many others.

Again, the BCS is estimated using a numeric scale that spans from 1 to 5. Intermediate scores have a precision of 0.25 (one increment). Nevertheless, cows with a BCS of less than 2.5 (very thin) and more than 4 (fat) present the same or very similar health problems, so precision is only required in the 2.5–4 range. Precision here means that the difference between the estimated BCS and the true BCS should not be greater than one or two increments. In Fig. 10.1 several examples of cows with different BCS values are shown. It can be seen that the cows with BCS of around 2 are very thin, whereas cows with 4 or more are fat. The cows displaying BCS scores between 3 and 4, present a more adequate body condition.

There are problems with the BCS that lead to the farms not using it on a regular basis [1]. To understand this, it is helpful to know how an expert estimates the BCS of a cow.

The parts where fat accumulates in cows are around the pelvis and the tail, thus the experts concentrate on the rear-end of the animals. Following a standardised method, they start looking at one side of a cow to determine the shape of the angle that spans from the hook bone, over the thurl to the pin bone. If this angle is V-shaped the BCS is 3 or less, if it is U-shaped, the BCS is above 3. This may be the most critical part, as the difference between V-shaped angles and U-shaped ones is not always clear and this difference determines the starting point of the evaluation. After that, the experts turn their attention to the rear of the cow, looking at it from behind. They evaluate the roundness of the hooks and how visible the sacral and tail-head ligaments are. They may also count the number of visible short ribs or touch some parts to get a feel of how much fat is covering them. Depending on the outcomes of these evaluations, the BCS is increased or decreased by a small amount after performing each one of them.

Thus it becomes clear that scoring a cow can be a very time consuming process. Assessors have to be instructed on body condition scoring to gain the knowledge required to rate animals. Even experts take up to 30 seconds to estimate a score. Since

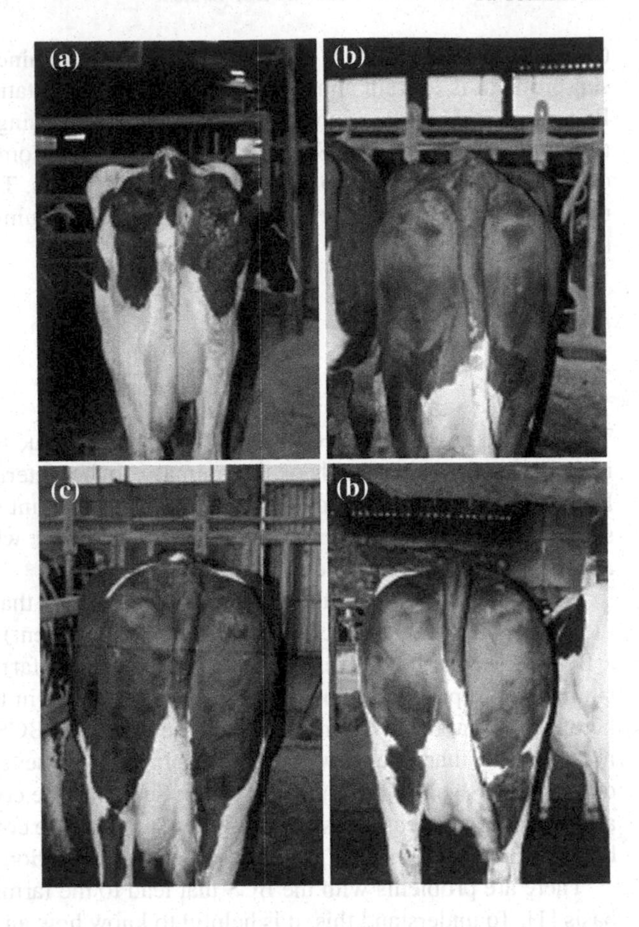

Fig. 10.1 Examples of cows with different BCS scores. **(a)** BCS=2 **(b)** BCS=3 **(c)** BCS=4 **(d)** BCS=5

some farms might consist of many cows the task can become impossible to perform in practice [3]. In fact, the BCS is also criticised due to the nature of the scoring process, which is very sensitive to subjective human judgements. The individual impressions of the experts might introduce inconsistencies leading to the scores not being comparable among different assessors [4]. There is therefore a need for an automatic procedure that can estimate the BCS of dairy cattle in a more objective and reliable way.

One of the first approaches on automatic body condition scoring from images for dairy cattle is that of Coffey et al. [5]. Some line patterns were painted on the animals using a red laser across the tail-head area of each cow. Those lines were then manually segmented from the images and quadratic curves were fitted. Those curves were successfully related to the BCS.

In Ferguson et al. [6] the authors tested the possibility of employing only images for body condition scoring. The experts scored the same cows, once live and then using the images. The results showed that no significant discrepancy between both scores could be found for the same expert. Indeed, those experts remarked that they could score using only the rear view of the cow. However, they could not successfully rate a cow with images of the side view. In this research it is therefore assumed that the back view holds more information for assessing the BCS.

Taking images from above, in Bewley et al. [7], the silhouettes of cows were manually segmented with 23 good recognisable anatomical points. From this shape, 15 angles could be calculated and a regression model was fitted to them. This model showed a very good correlation with the BCS.

Using the same view, but employing thermal cameras, which simplifies the segmentation of the contour, in Halachmi et al. [8] the back of the cow was fitted to a parabola. Fatter cows should have a rounder profile whereas thinner cows should show a significant deviation from the parabola model. The authors demonstrated indeed that the goodness of fit is a good indicator of the BCS.

More recently in Azzaro et al. [9] the same points of Bewley et al. [7] are also manually determined. But in contrast to the other work, in Azzaro et al. [9] the shapes are used for building a point distribution model applying statistical shape analysis. In this model each shape is described by a parameter vector that is much smaller than the shape itself. These vectors constitute the input for a regression model that achieves results similar to those of Bewley et al. [7] and Halachmi et al. [8].

10.3 Assessing the Error Between Experts and Constructing a BCS Dataset

As a first step towards building the ABiCA system it is necessary to assess the feasibility of producing an automatic body condition scoring system using images as well as to determine what error levels would be acceptable in terms of those of the human experts. In this line, several experiments were carried out. The experiments involved two experts and visits to different farms, where sets of images of the back view of different cows were taken. With the scores of the experts and these images, a dataset was constructed for training and evaluating the body condition scoring system.

10.3.1 Reliability of the Experts

The reliability of the scores provided by the experts causes some concerns since they use subjective visual information. Their scores could therefore be inconsistent. Indeed, an expert may rate the same cow with different scores even when its BCS has not changed.

In a first test, the two experts (*Exp1* and *Exp2* from here on) rated more than 30 cows (both scored the same cows) on a farm on two different days. This type of scores are referenced in the text from now on as *in-situ*, since they have been estimated by evaluating a cow on the farm. The lapse of time between the two days of scoring was of three weeks, an interval where the BCS, according to the experts, can be regarded more or less constant. The scores from each expert were recorded and a picture of the back view of the rated cows was catalogued.[2]

As a second test, those pictures were then presented to the experts weeks after the first test, to exclude that they had memorised some cow. The experts scored then the cows on the pictures to see if there were any inconsistencies when scoring using images.

In the first test, for each expert, the Mean Absolute Error (MAE) between the *in-situ* scores in different days was calculated. In the second test, for each expert, the MAE between the scores over images and those *in-situ* was calculated. Since a single value might not be representative enough in order to compare results, an interval for each MAE was also calculated using bootstrapping, so that the true values were found within that interval with 90 % probability. The results can be seen in Fig. 10.2. In this figure, the results of the first test are labelled as "*in situ*" for each of the experts. They can be taken as a measure of how consistent the estimations of an expert are. Experts with more consistent scores are preferred, since this reduces the spurious variability of the data. The results for the second test are labelled as "images vs. *in-situ*".

Two conclusions can be extracted from Fig. 10.2. First of all there does not seem to be any significant difference on how consistent the estimations of the experts are, so any of the two could be chosen as a reference for training the body condition scoring system. Nevertheless, the expert that appeared to show less variability (smaller interval) was chosen. In this case it was Exp1. The second conclusion is that the

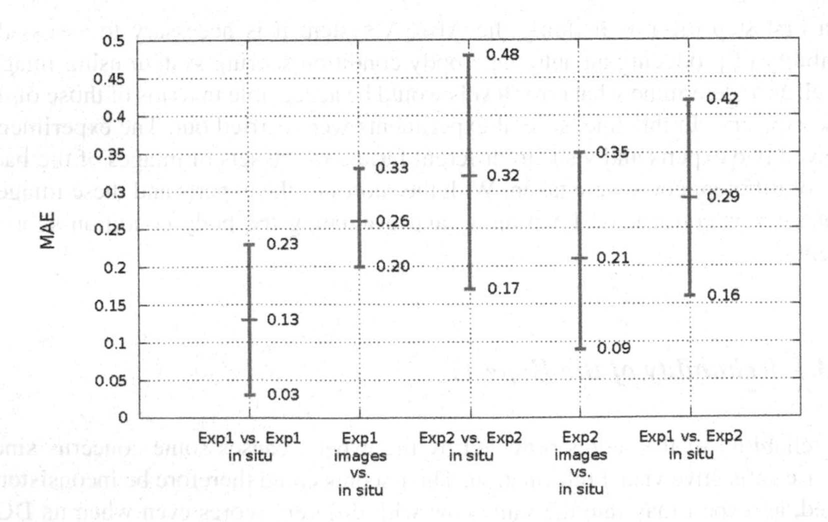

Fig. 10.2 Errors measured for the experts. The central value of the intervals is the MAE value obtained from the sample. The extremes were calculated using bootstrapping

[2] The pictures were shot with a Canon 300D consumer camera with its kit lens on it. Given the poor illumination in the farms, most of the pictures used the built-in flash of the camera.

error of scoring over images with respect to the reference values (the scores *in-situ*) is similar to the error of scoring a cow in different moments (but with constant BCS) *in-situ*. Thus the discrepancy between scoring *in-situ* and over images is well within the allowed scoring error of two increments.[3] This in turn means that a system that relies only on images from the back view of cows for body condition scoring should be feasible, as there is enough information on that view. This result confirms the original premise of Ferguson et al. [6].

10.3.2 Dataset and Error Estimation

Again, during the experiments with the experts each BCS scored by them was recorded and images were taken of the back view of the cow. This allowed the construction of a dataset of labelled pictures with their corresponding BCS that could be used for training a body condition scoring system. The dataset contains 125 images of back views of cows from different farms. The scores assigned to each cow of the dataset were those of Exp1, chosen as the reference expert. Note that not all cows were evaluated by the two experts. Figure 10.3 shows the frequency of each BCS value in the dataset.

The figure shows a typical situation of many datasets in machine learning. Most of the values are in the central range of the scale, in this case 2.5–4. Values on the extremes are less likely to be observed, so they are not well represented. No cow with a BCS of less than 2 was recorded, since those are cases of animals that are too

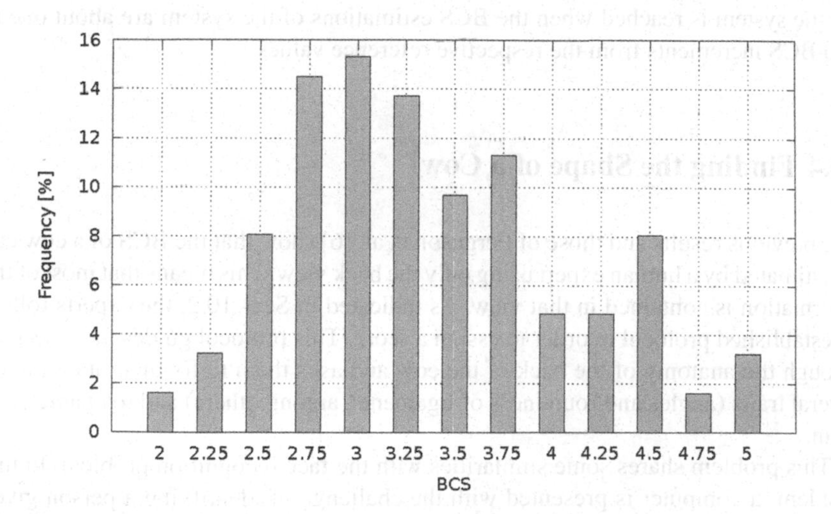

Fig. 10.3 Frequency of each BCS in the dataset compiled for training the system

[3] A scoring error of 0.5.

emaciated to live for a long time. Consequently the whole BCS scale could not be completed.

These circumstances make the task of training an automatic system with those data challenging. The data could be balanced, but this would almost certainly cut out too many values in the central range. The only real alternative in such cases is to substantially increase the sample size, but this was not possible in this case. However, as mentioned in Sect. 10.2, where precision is really needed is just in the well represented 2.5–4 range, whereas values on the extremes are less important. Thus this dataset fits the purposes of the system.

On the other hand, in order to assess the quality of the body condition scoring system results, the error that the system could assume had to be estimated. Ideally the results should be compared to the true BCS. The problem is that there is not a single value that can be regarded as the true value, since both scores of the experts deserve the same credibility. The ground truth has thus some uncertainty. One possibility to overcome this problem, is to compare the error of the system to the expected error difference between two human experts. Similar error outcomes would mean that the system has an acceptable performance. In this line, the error between Exp1 and Exp2 was calculated with the available data as the MAE. An interval where the true error should be with high probability (90 %) was also calculated. Figure 10.2 shows the error labelled as "Exp1 vs. Exp2 in *in-situ*".

Taking the error of the experts into account, the error that could be assumed for the system is about 0.25 (one increment) of the reference BCS. However, as indicated in Sect. 10.3.1, the scores of the experts suffer from an uncertainty of about another BCS increment too. Thus a more realistic approach for the assessment of the system's error should consider these two uncertainties. This means that an acceptable performance for the system is reached when the BCS estimations of the system are about one to two BCS increments from the respective reference values.

10.4 Finding the Shape of a Cow

The previous results and those of Ferguson et al. [6] show that the BCS of a cow can be estimated by a human expert using only the back view. This means that most of the information is contained in that view. As indicated in Sect. 10.2, the experts follow an established protocol in order to assign a score. This protocol guides the assessors through the anatomy of the back of the cow and asks them to fix their attention on several traits (angles and roundness of ligaments, among others) and to characterise them.

This problem shares some similarities with the face recognition problem. In this problem, a computer is presented with the challenge of identifying a person given a picture of a face. There are several traits that can be used to accomplish the task and a popular way to do it is to model the face using its shape so that these relevant measures can be easily calculated. A widely employed technique for obtaining the shape are ASMs [2].

The body condition scoring system also needs a representation of the back of the cow that simplifies the measurement of the features that are important in order to be able to estimate the BCS. Following the same approach as for the face recognition problem, the back of the cow is modelled using its shape. For finding the shape ASMs are employed.

10.4.1 Active Shape Models

ASMs were introduced by Cootes and Taylor [2]. Their use requires four different components: a training set of images annotated with shapes, a global shape model, a local appearance model for each point of a shape (also called landmarks) and a search method.

The global shape model or (PDM) is built by statistical shape analysis of a training set of shapes. A shape of n points is defined as a vector $(x_1, \ldots, x_n, y_1, \ldots, y_n)^T$. In this notation, x_i is the x coordinate of the i-th point of the shape, whereas y_i is its y coordinate. The goal is to model the variability of the shapes by a simple linear model that takes the variability of the shapes into account.

Usually, the shapes are first normalised so as to have the same scale and aligned, so that all share a common co-ordinate frame, but neither of these steps is really a requirement. For building the model, the mean shape $\bar{\mathbf{x}}$ is calculated and the principal directions of the different variations present in the training set of shapes are found by means of a principal component analysis. However, not all principal components are preserved, only those that account for an admissible amount of variability. Typical values for this amount are 95 % to 98 %. The global shape model is expressed in Eq. 10.1, where $\mathbf{P}$ is the matrix of the selected principal components, whereas $\mathbf{b}$ is a vector with values that are called shape parameters.

$$\mathbf{x} \approx \bar{\mathbf{x}} + \mathbf{P}\mathbf{b} \tag{10.1}$$

To guide the search, for each point of the shape, a local appearance model is built. These models try to capture the local structure around each landmark as well as possible. As for the local structure, normally linear profiles orthogonal to the shape boundary are used. Each of those profiles is built by sampling the grey level values of k pixels at each side of a landmark. That is, the profiles have a length of $2k+1$ pixels.

The simplest local appearance model is searching for a nearby edge to the current landmark. This model has the great advantage that it does not require any training, however, it often gets trapped by other borders.

A very common alternative to the edge search is using the Mahalanobis local appearance model as in Cootes and Taylor [10]. For this model, normalised first derivatives of the profiles $\mathbf{g}_i \ldots \mathbf{g}_s$ at the same landmark of the s shapes are computed. By calculating the mean $\bar{\mathbf{g}}$ and the co-variance $\mathbf{S}_g$ of those profiles and assuming that they follow a multivariate Gaussian distribution, the Mahalanobis distance of Eq. 10.2

from a new profile $\mathbf{g}_{new}$ to the sampled profiles can be computed.

$$d(\mathbf{g}_{new}) = (\mathbf{g}_{new} - \bar{\mathbf{g}})\, \mathbf{S}_g^{-1}\, (\mathbf{g}_{new} - \bar{\mathbf{g}}) \tag{10.2}$$

This distance is related to the probability that the new profile belongs to the original distribution of the sampled profiles. Minimising the distance, maximises this probability. It is important to note that, since the inverse of the co-variance matrix is needed, this distance can not be computed in case of a singular matrix. This means that the Mahalanobis local appearance model can not always be used.

Again, these appearance models are used during the search for estimating the new positions of the landmarks. The search begins with an initial shape estimate. In each iteration and for each landmark a region of size $2n_s + 1$ normal to the landmark is sampled by moving it $n_s - k$ positions at both sides of its original location. The local appearance models are used to estimate the best displacement and the shape is deformed accordingly. The global shape model of Eq. 10.1 is then applied to ensure that the shape is still a valid shape and conforms to the variations seen on the training set. The model is applied by calculating the shape parameters **b** and constraining them, for example, by ensuring that $|b_i| \leq 3\sqrt{\lambda_i}$ iteratively. Here λ_i is the eigenvalue of the i-th principal component. Since it can be shown that each eigenvalue is the same as the variance of its corresponding eigenvector (principal component), by limiting b_i to a multiple of $\sqrt{\lambda_i}$ a percentage of the variability of the training set is covered. The search repeats this process for a fixed number of times N_{max} or until a specific stop condition is met.

Since there is no guarantee that this method converges and it might even be the case that the search gets stuck in local optima, sometimes a multi-resolution framework is employed. The search begins at the coarsest resolution, runs a number of iterations in that level or until a convergence condition has been fulfilled and then the search continues in the next level.

10.4.2 Experimenting with More Local Appearance Models

The pictures that serve as input for the system are taken using hand-held cameras in very challenging environments. Figure 10.4 shows examples of the pictures that were shot. Several problems can be observed on these images. The space behind the cows is not always the same, therefore the scale of the cows in the pictures is variable. Notice also the huge illumination variability present in the farms. Sometimes the built-in flash of the camera had to be used, in other situations a sufficiently strong light source prevented it from popping up. The flash might also cast shadows to nearby walls or columns. There are pictures with very dark backgrounds, whereas there are others burned with highlights. Others show transitions from sunny sides to shadow areas. The backgrounds are very cluttered with whatever farm tools or even other cows. Finally, even the skin of the cows is non-uniform as it contains spots that

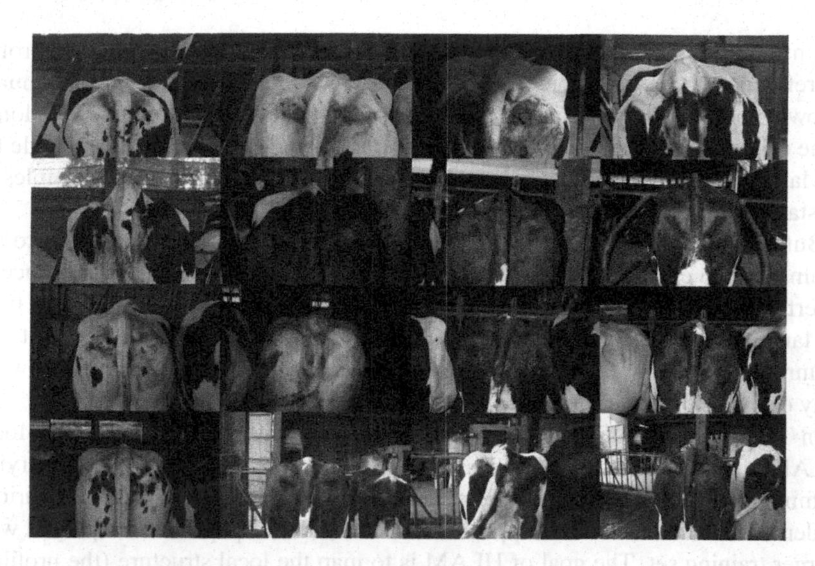

Fig. 10.4 Examples of the pictures taken on several farms

introduce spurious borders. In practice this breaks the contour of a cow and acts like camouflage.

In such situations the search in an ASM depends to a great extent on the ability of the local appearance models to describe the local structure accurately. As commented in Sect. 10.4.1, the simplest local appearance model is searching for strong edges in the neighbourhood of the current landmarks. The advantage of this approach is that no training needs to be carried out and the memory requirements during search are low. Nevertheless, searching for nearby borders might not be suitable at all when spurious borders are present, as is the case of some images, since the search can be easily fooled by such borders.

The use of the Mahalanobis appearance model is also very straightforward, but it requires more resources than the edge search since a multivariate Gaussian model for each landmark has to be trained. On the other hand it can perform better because it might model the structure around landmarks more accurately. However this model is only useful when the local structures of the sampled profiles follow the assumed distribution. Since the inverse of the co-variance matrix is needed, the model can be built only if this matrix is invertible or not heavily ill-conditioned. For training sets with few samples this cannot always be ensured. Nonetheless, the training set should be large enough. Even so, it is not clear that the local structure of the profiles can be modelled as a multi-dimensional Gaussian distribution.

In this chapter several local appearance models were tested. One of the main issues detected on the pictures is the non-uniform illumination. To somehow overcome this problem, a new local appearance model based on the spectral angle of the sampled profiles was developed. Like for the Mahalanobis appearance model, profiles of length $2k+1$ are sampled normal to the shape contour at each landmark. But instead of computing the first derivatives of these profiles, the spectral angle from each profile to a reference profile is computed. If it is assumed that the angles for a given landmark follow a normal distribution, then a measure of how likely it is that a profile belongs to the original distribution is to compute how far its angle is from the mean angle for that landmark. The distance can be calculated for instance in terms of multiples of the standard deviation of the angles. This is a very simple and fast model.

But assuming a fixed shape for the distribution of the local structure around landmarks and estimating its parameters might also not be suitable. For instance, it is perfectly possible for the type of distribution of the local structure to vary from one landmark to another. Describing the local structure with high fidelity might also be unnecessary. In this case, a model that guides the search by estimating how far away each position is from the true position could be used.

In Tedín et al. [11] the authors proposed the Heuristic Local Appearance Model (HLAM). They successfully applied it for a small training set of hands under varying illumination and for a synthetic training set where the images presented spurious borders that fooled other local appearance models. In this paper it is employed with a larger training set. The goal of HLAM is to map the local structure (the profiles) to a custom distance function. In a certain way, the distribution of the local structure is imposed by a known function. This function can also be regarded as the fitness of a profile as in Eq. 10.3.

$$fitness\ (profile) = 1 - e^{-\left(\frac{\delta}{n_s - k}\right)^2} \tag{10.3}$$

The problem can be thus treated as a classification problem. During training a separate classifier for each landmark is trained by sampling a region n_s pixels either side of the current landmark. Since a profile has size $2k + 1$, there are $2(n_s - k) + 1$ profiles for training. The fitness of a profile is dependent on the displacement δ of the central pixel of the profile to the true landmark. When searching, the classifier is used to estimate the fitness of a profile.

Any classifier could be used. In this case, the M5' classifier [12] was chosen, as it performed well in Tedín et al. [11]. The implementation of the classifier used in this work is the one found in the well-known WEKA [13] data mining software distribution.

Nevertheless, there is no *a priori* information to help to decide which of these local appearance models is most suited to the problem. Thus experiments with all of them were performed. In Sect. 10.6.1 the results for each of the local appearance models are discussed.

10.4.3 Extension to Multi-Channel Images

Finding the back shape of a cow in images taken using hand-held cameras with uncontrolled illumination and backgrounds can be challenging. Thus it could be very useful to make use of more information than that provided by the grey level values of the pixels. For dealing with colour images, an extension to the local appearance models that can handle more than one component per pixel is required.

Two straightforward approaches can be followed: train a local appearance model for each component or train a local appearance model using all the components at once. The simplest one is to apply a local appearance model to each one of the components and take a reasonable value computed from the individual outcomes of each one of them. This value could be for instance the mean of the outcomes of Eq. 10.2 for the Mahalanobis appearance model or the maximum gradient change (the strongest border) on any component for the edge search appearance model. Other approaches such as taking the median in order to eliminate outliers as in Koschan et al. [14] or a multi-objective method could be used.

This approach has the downside of requiring more resources than using the grey level values, since for each landmark and, additionally, each component a separate local appearance model has to be trained (for those that require training). However it is very easy to implement and requires little modification of the existing implementations that use the grey levels. Thus this is the procedure followed in this work for all appearance models. For the edge search appearance model a search for the strongest edge on any component is applied. For the other models, the mean of the individual outcomes of each component is taken. This method is called a "monochromatic-based technique" in Koschan et al. [14] and "component form" in Tedín et al. [11] as it uses the different channels separately.

A second method for dealing with multi-channel images could also be applied. Instead of having a separate appearance model for each component, all of the components can be concatenated into a single feature vector. Then only a single appearance model for each landmark is needed. This scheme has access to the whole information of a pixel at once and it might find useful connections between the components. In Tedín et al. [11] this approach is called "compound form" as all of the information from the channels is used as whole. This procedure is implemented in this paper only for HLAM. There are thus two versions of HLAM: a component form and a compound form.

10.4.4 The Back Shape of a Cow

For the ASMs to work, a training set of labelled images with their respective shapes is required. This involves annotating each image using many points so that the shape is adequately described. If the training set contains a large number of images, this task can be extremely time consuming and error prone.

Therefore a semi-automatic mechanism was used to label the images and to thus prepare the final training set for the ASMs. Each image of the back of a cow was first cropped and scaled to a resolution of 255×170 pixels. This saves computer resources and somewhat simplifies the problem of finding the shape. In a real set-up this step should be automatic, for instance by searching for the pose (scale, orientation and position) of the back of the cow. Again this is similar to the face recognition problem where first a coarse approximation of the face pose is needed. A common method for this case is the Viola and Jones face detector [15]. To find such a pose of the back of the cow automatically is a matter of future work. Thus it is assumed that the pose has already been found.

Instead of labelling the images by means of every point of the shape, each shape is modelled as a fourth degree spline of Bezier curves. Thus only the control points of the Bezier curves are needed and the rest of the points (which are called secondary landmarks) can be generated automatically by sampling the spline. Figure 10.5 displays a schematic of the back shape of a cow.

The secondary landmarks are, obviously, less reliable than the main landmarks, so it is advisable to put some of the control points of the Bezier curves on main landmarks. This way at least a few of these points can serve as a reference for detailed measurements.

As seen in Fig. 10.5, the back shape of the cow is a composition of the most prominent lines of the back. The top line is normally very visible, except for very cluttered backgrounds, poor illumination or when the spots of the skin of the cow break its contour. The other line is generally more difficult to see, especially for fatter cows where it can almost vanish. The spots can also cause trouble when finding this line.

In Fig. 10.6 several examples of ASM searches are shown for the sake of illustrating the search results using the rear-end shape. In this case the simple edge appearance model has been used as the local model for the ASM. Since the upper shape is normally a hard border, it is generally easier to find, while the other sub-shape is more difficult for this type of ASM.

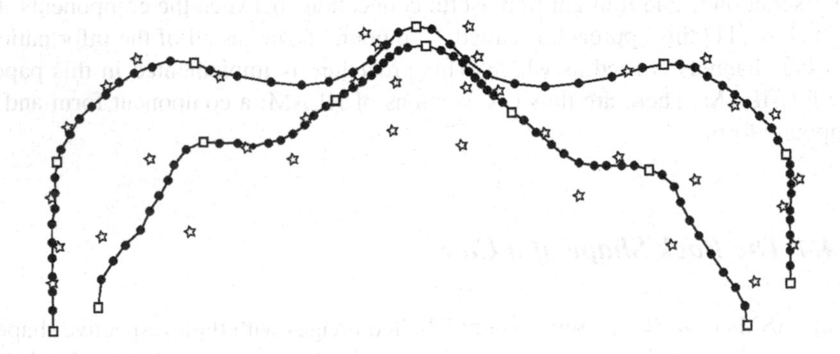

Fig. 10.5 Schematic of the back shape of a cow. The control points are depicted as stars, the main landmarks as squares and the secondary landmarks as solid dots

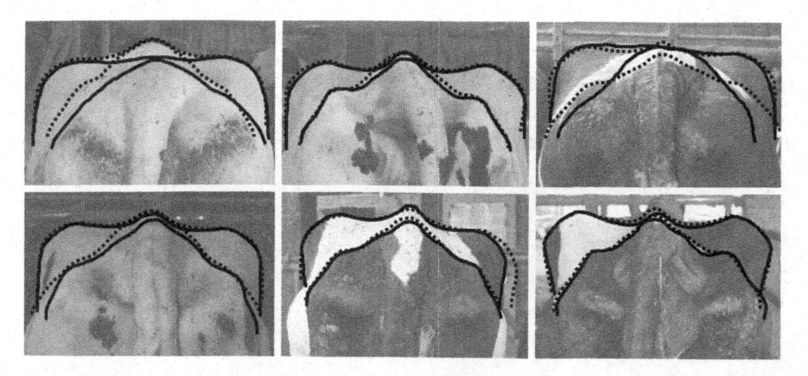

Fig. 10.6 Examples of ASM searches. The found shape is the dashed line whereas the expected is the solid one

10.5 Using the Shape of the Cow for Obtaining the BCS

Once a shape has been found the system has to estimate the BCS of the cow from it. Previous papers extracted features from the images and tried to fit a regression model to those values. In Bewley et al. [7] 15 angles from the contour of the cow were used. On the other hand the shape parameters of an ASM were successfully employed in Azzaro et al. [9]. In these cases the contour was segmented using images from the top view of the animals.

In this work however the same view as that of the experts is employed (the rear-end view). Thus, there is no access to those features. Nevertheless, taking the angles between segments as features could also be a good approach for the back view. This hypothesis was tested by measuring 23 angles of the back view of the cow and by training machine learning classifiers using them. The details are presented in Sect. 10.5.1. Additionally, in Sect. 10.5.2 more general features were explored. Those features were found and combined by means of symbolic regression using genetic programming [16]. Both of these methods represent a machine learning approach to the problem in contrast to the more statistically based methods of Bewley et al. and Azzaro et al. [7, 9]. Section 10.5.3 compares the results of the two methods.

10.5.1 BCS Estimation by Classifiers

For estimating the BCS from shapes of the back view of cows, experiments with six machine learning classifiers implemented in WEKA [13] were performed. The experiments consisted in cross-validation tests using 23 angles extracted from the 125 shapes of the training set compiled in Sect. 10.4. A schematic of the angles used as features can be seen in Fig. 10.7.

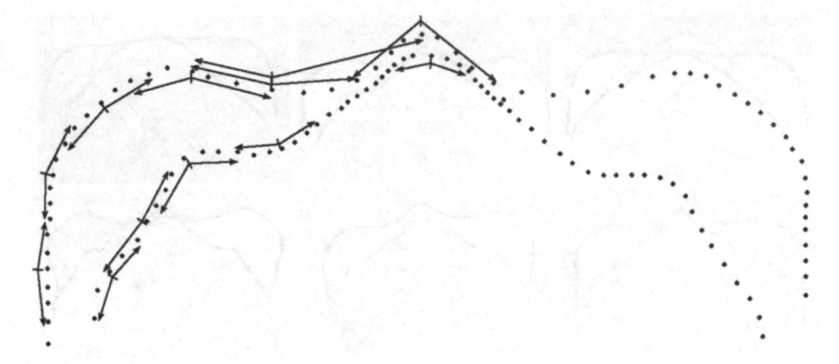

Fig. 10.7 Schematic of the angles used as features for the machine learning classifiers over a cow's back shape. Only the angles on one side are shown

The classifiers that were tested were: M5P, GaussianProcesses, Ibk (with $k = 1$ and $k = 2$), KStar and MLP. Those are the names of the Java classes that implement these classifiers. This nomenclature has been respected so that the reader can consult the WEKA documentation for details on the algorithms. All classifiers where applied with the default set of WEKA parameters for each method. Additionally the ZeroR classifier was employed. This is a special classifier that just returns the mean of the target values of the training set. It has thus no predictive value, but it serves well as a reference performance for the other classifiers. The output of the classifiers with real-valued outcome was systematically rounded to the nearest multiple of 0.25, so that valid BCS scores where always generated.

10.5.2 BCS Estimation by Symbolic Regression

In this section an approach to the estimation of the BCS is presented that uses more features than those of Bewley et al. and Azzaro et al. [7, 9] and Sect. 10.5.1.

With the help of the JCLEC library,[4] a symbolic regression model of the BCS was evolved by means of genetic programming. Symbolic regression tries to approximate the target values (the BCS in this case) by applying a combination of any kind of functions to the input data. It is thus a more general approach than for example a polynomial regression where the relationship between the inputs and the targets are represented by an n-th degree polynomial. On the other hand, genetic programming is an evolutionary computation tool to automatically find computer programs or functions.

In order to be able to use genetic programming a set of terminals and functionals need to be defined. Terminals are constant expressions and functionals are functions

[4] See http://jclec.sourceforge.net. Last access March 2013.

that are applied to the terminals and to the result of other functionals in order to compute a final result.

The set of terminals employed here were a subset of the points of the back shape of a cow, real valued constants such as π or 1, random real valued constants and the two boolean constants *true* and *false*.

However, the set of functionals is somewhat richer, since they serve two purposes. First of all, the functionals apply functions such as tangent or cosine over their input values. On the other hand, they are also responsible for extracting the features from the shape, for example by calculating the angle spanned by three points.

Examples of functionals that only apply a function to its input are the *cosine*, *tangent*, *sine*, *pow*, *square root*, *max*, *min* and the arithmetic operators $+$, $-$, $*$ and $/$. Defined logical functionals are *and*, *or* and *not*, the comparison operators $<, >, \leq$ and $\geq$ and an *if...then...else* operator. A number of functionals that extract features from the shapes such as distances between two points, angles and measures of curvature were also implemented.

After executing a series of experiments, a sufficiently good function was found with a rather compact form that uses only two angles to approximate the BCS of a cow given its back shape. This function is called *bcsSymb* and is shown in Eq. (10.4. Note that the outcome of *bcsSymb* must be rounded to the nearest multiple of 0.25 to get a valid BCS. The two angles that are used can be seen in Fig. 10.8. The experts look at many more anatomical areas of the cow in order to assess a BCS, but it seems that these two angles hold enough information for scoring a cow.

$$\begin{aligned}
bcsSymb\,(\alpha, \beta) &= 0.01 \cdot e^{cos(tan(\alpha)) - f(\beta) - D} \\
f\,(\beta) &= A \cdot sin\left(B \cdot max\left(C, tan\,(E \cdot \beta)^{F}\right)^{-1}\right)^{-1} \\
A &= 0.8672154756813469 \\
B &= -0.08471845014 \\
C &= 0.4797669949635008 \\
D &= 0.11905455891510185 \\
E &= 0.9119892133700969 \\
F &= 0.831475415459637
\end{aligned} \tag{10.4}$$

10.5.3 Choice of the Estimation Method

In this section the performance of the classifiers and *bcsSymb* is compared. The measure for comparing the results is the MAE of the predicted BCS values with respect to the expected values. As in Sect. 10.3, an interval for this error was computed using bootstrapping. The results of the classifiers and the *bcsSymb* function are shown in Fig. 10.9.

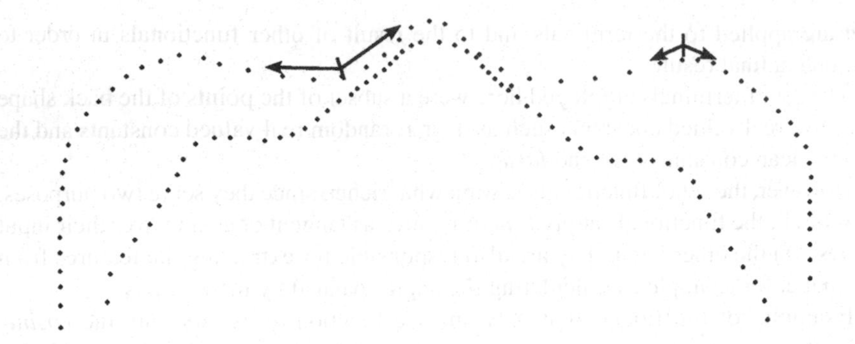

Fig. 10.8 The two angles that are sufficient for estimating the BCS with the function obtained by symbolic regression. The angle on the left is α, the one on the right is β

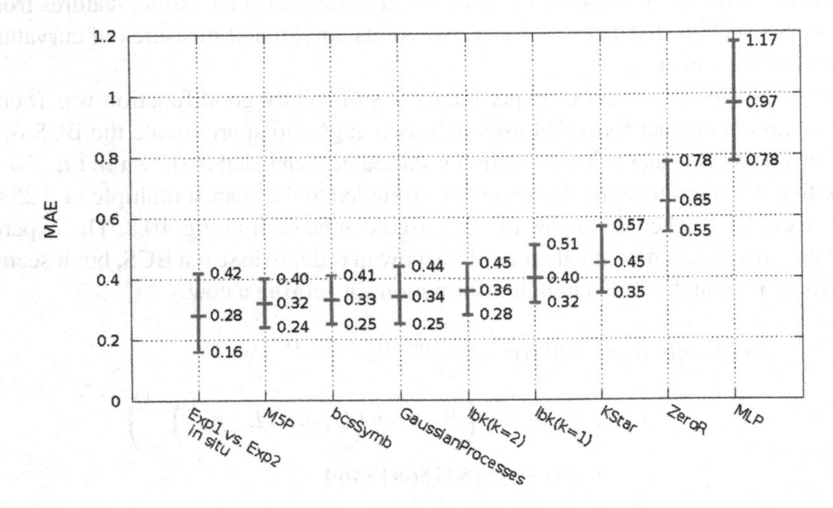

Fig. 10.9 Results of the classifiers and the *bcsSymb* function compared to the error between experts

As shown in the figure, all the classifiers and the *bcsSymb* function have learned the relationship between a shape and its BCS to some extent, since their results differ from those of ZeroR. However, the results for KStar are not so clear. On the other hand, MLP shows a really poor performance, which might be related to the fact that the standard parameters of WEKA were used. A different set-up could perhaps improve those results.

What is more, either the M5P, GaussianProcesses, or Ibk classifiers or the *bcsSymb* function can be used for a reliable body condition scoring system, considering that their errors are very similar to the error between experts. The *bcsSymb* function was chosen, as it presents the smallest error variability, has a relative simple mathematical closed form that is easily transferable to a computer program and only uses two angles in a zone that is more accessible to segmentation algorithms than other parts of the cow backside.

10.6 The ABiCA Final System

In Sect. 10.4 an outline of how to describe the rear-end view of the cow using its shape was described. Given that shape, Sect. 10.5 showed how to estimate the BCS of the cow from it. Thus the first two sub-problems formulated in Sect. 10.1 were addressed independently. In this section the solutions of these two sub-problems are combined to construct a complete body condition scoring system from images of the back view of cows. This system is called ABiCA.

For the estimation of the BCS to work with the *bcsSymb* function, the two angles that are required as input must be measured. This in turn requires finding the shape of the back of the cow. As ASMs have been chosen for this task, the ASM that allows to locate those angles with the most precision must be found.

However, ASMs have quite a few parameters. For assessing the performance of the local appearance models defined in Sect. 10.4.2, the best ASM for each of them must be found. Since this is a difficult task to complete by hand, an EA is used to automatically tune the parameters of the ASMs for each one of the appearance models. The DE algorithm [17] has been chosen because it has proven to be a good contender [18] for different kinds of challenging problems.

The parameters of the ASMs that are tuned by the EA were k, n_s, N_{max}, the amount of variability that must be covered by the principal components of the shape model or the colour space used (Monochrome, RGB, HSI) and among others. Five different experiments were performed, each one with a different local appearance models from Sects. 10.4.2 and 10.4.3. Those appearance models were the edge search procedure, the Mahalanobis appearance model, the spectral angle appearance model and both forms of HLAM (component and compound).

On the other hand, as every other EA, DE needs an objective function that assigns a fitness value to each individual (candidate solution) of the population. In this case an individual is a combination of values for the parameters that must be tuned for each ASM with a particular local appearance model. The performance of an ASM can be measured assessing how close the shape found is to the expected shape. Therefore, in each evaluation of an individual, a cross-validation process is performed that calculates a measure of the distance from the shape found after running an ASM search to the expected one. The set of images and shapes used were those compiled in Sect. 10.4.4. An evaluation might take minutes and many evaluations must be carried out. Thus the evolutionary computation library of Caamaño et al. [19] and its support for parallelizing the execution of the algorithm has been used.

As the measure of distance between the found and expected shapes, the point to point distance could be used. However, this distance can be too dependent on the secondary landmarks. Remember that those are landmarks generated automatically by sampling the Bezier spline that defines the shape. If the secondary landmarks suffer a shift along the true shape, the point to point distance will detect this movement. Actually this shift does not matter if one wants to measure angles, as is the case. Hence the point to point distance would probably put too much evolutionary pressure and the problem of tuning the ASM parameters would be even more challenging.

Consequently, the measure that was adopted to compare two shapes was that of Eq. 10.5.

$$\delta = \frac{d\,(e, o)}{A\,(e)} \tag{10.5}$$

The $d\,(e, o)$ is the difference in pixels between the area of the expected shape e and the area of the obtained shape o, whereas $A\,(e)$ is the area of e. If a shape is an open path, the path is closed for the sake of computing the area. On the other hand, if a shape is made up of several sub-shapes, as is usually the case here, then Eq. 10.6 is used. Where e_i and o_i are, respectively, the i-th expected and obtained sub-shapes.

$$\delta = \frac{\sum_i^n d\,(e_i, o_i)}{\sum_i^n A\,(e_i)} \tag{10.6}$$

10.6.1 Results

After obtaining the best set of parameters for each ASM, the mean absolute error was calculated between the expected BCS scores and those estimated by *bcsSymb*. The angles for *bcsSymb* were measured from the shapes obtained by the ASM searches. Again, intervals for these errors were constructed. The results can be observed in Fig. 10.10 compared to the error between experts.

Because of the scores from two different experts, there is not an unique true value for the BCS of the cows. The error between experts serves thus as a reference for the error of the system. This means that the system should have an error of about 0.25 (one BCS increment) to be considered comparable to an expert. However, since an expert

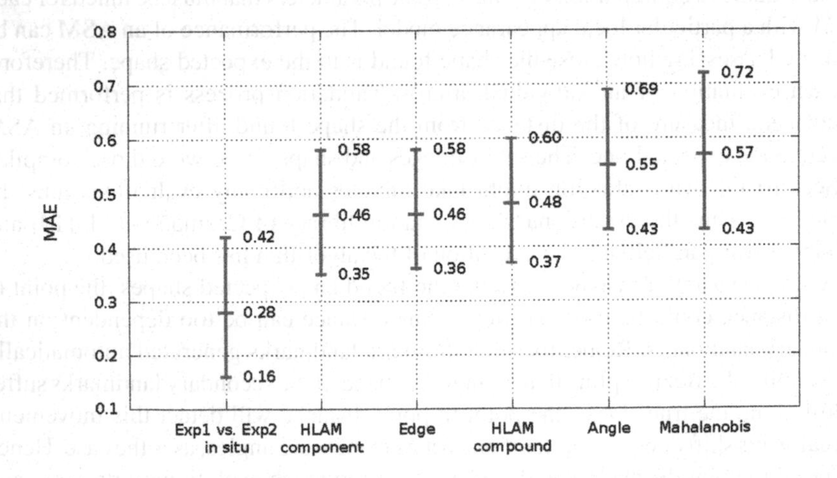

Fig. 10.10 Errors of the evolved ASMs compared to the error between experts

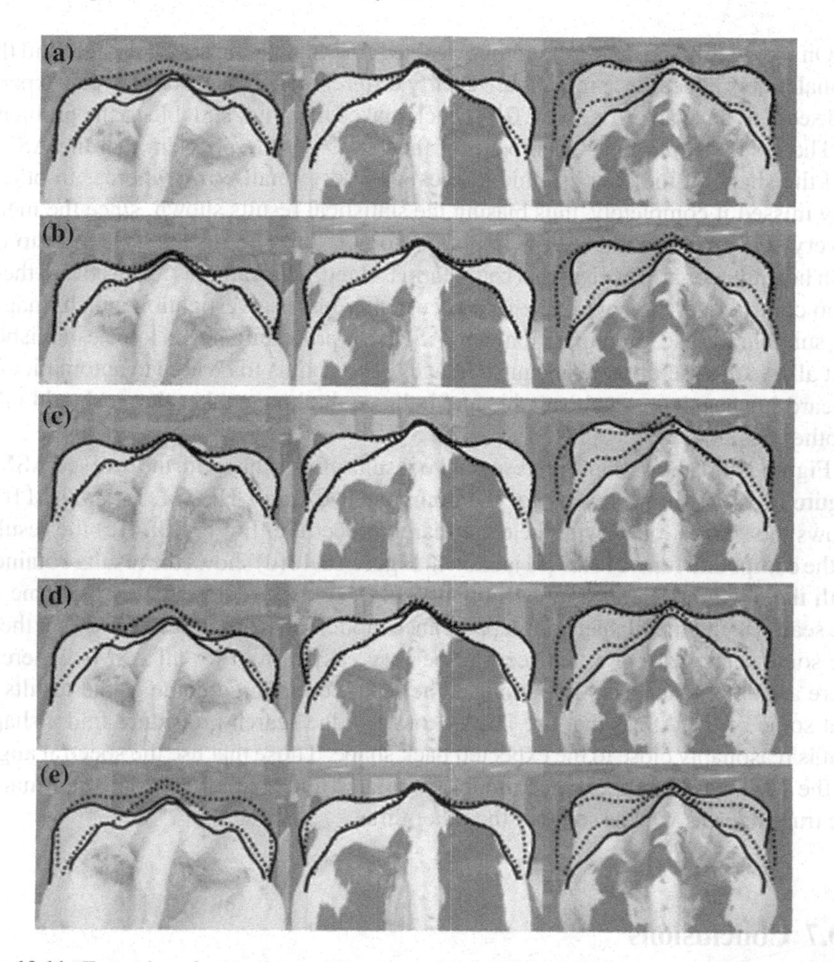

Fig. 10.11 Examples of executions with each evolved ASM. The found shape is shown dashed whilst the expected one is shown as a solid line. (**a**) HLAM component (**b**) Edge (**c**) HLAM compound (**d**) Angle (**e**) Mahalanobis

shows itself an additional uncertainty of about another increment when scoring over images, it is considered that the system offers a reasonable performance for an error in the range of 0.25 to 0.5. The evolved ASMs with HLAM (both component and compound forms) and the one that uses the edge search appearance model are in this case quite acceptable. Indeed these three appearance models perform very similarly. The spread of the intervals make them even comparable to the error between experts and there is clear room for improvement. The behaviour of the ASM with the edge search appearance model is interesting. Despite being a very simple model, it is on the same level as both HLAMs. This is explained by the fact that the upper line of the back shape often lies on a relative hard edge and it is thus easy to find. Since *bcsSymb* uses angles measured over this line, the results are not surprisingly quite good.

On the other hand the errors for the evolved ASMs with the spectral angle and the Mahalanobis appearance models are clearly different from the error between experts and seem to have an error above 0.5, which makes them not suitable to the problem.

The results could have been better if there was not the problem that the ASMs find the shape of the cow in some images with very small error, whereas in others they missed it completely, thus biasing the statistical results shown, since the mean is very sensitive to outliers. It is important to note that the results have a lot to do with how images are taken in real cattle farms. Hand-held cameras are used and there is no control over the environment. It is currently under investigation which images are suitable for use in conjunction with ASMs. A procedure needs to be established that allows the use of more adequate images. The goal is to be able to automatically discard images that are not suitable and indicate the image taker that it should take another one.

Figure 10.11 shows some representative results of searches with the evolved ASMs. Figure 10.11(a) shows results for the component form of HLAM. Figure 10.11(b) shows the shapes found with the edge search procedure. In Fig. 10.11(c) the results of the compound form of HLAM is shown. Figure 10.11(d) shows the results obtained with the angle appearance model. Finally, Fig. 10.11(e) represents the outcome of the search using the Mahalanobis appearance model. It is important to note that there are some shapes of cows that seem to be very easy to find for all ASMs, whereas there are others that seem impossible. The most common outcome of the results is that some of the ASMs that use HLAM or the edge search procedure find a shape that is reasonably close to the expected back shape. Those that use the spectral angle or the Mahalanobis appearance models normally find a result that does not match the true shape as well as some of the other three.

10.7 Conclusions

This chapter describes the steps in the procedure that has been followed to build ABiCA, a body condition scoring system from images of the back view of dairy cattle obtained by farmers using hand held cameras. This is quite an important problem as this type of scoring allows the farmers and veterinarians to improve the management of their farms.

After introducing the BCS and its relevance to farms, a review of previous papers on the matter was carried out and a description of how experts score a cow was provided. Body condition scoring is a visual, subjective task and the farms would benefit from a system that can estimate this score in an automatic and objective fashion.

The problem was divided into three goals. First, a computationally usable representation of the cow had to be chosen and this representation had to be found in an image. Second, this representation had to serve as input for a procedure that could estimate the BCS of the cow from it. Finally, the two aforementioned methods had to be combined to build ABiCA.

The rear-end shape of the cows was chosen as the computationally tractable representation to use, since it holds the most information about the BCS. A procedure using Active Shape Models was described in order to find this shape within the images. During visits to different farms a set of pictures of the cows were shot and their BCS were registered from the information provided by two experts. Those images were processed and labelled with the true shapes of the cows so that they could be used as a training set for the ASM. Additionally, several experiments with the two experts allowed the selection of one of them as the reference expert and the estimation of the error between them. This error is used for assessing the performance of the system.

Two approaches that contrast with the methods followed by other authors were tested in order to estimate the BCS from the back shape of cows. The first approach consisted in measuring 23 angles from the back shape of the cow and in estimating the BCS using these angles as input features for machine learning classifiers. The second approach, on the other hand, addressed the problem through the use of symbolic regression by means of genetic programming to automatically select the relevant features of the cow and to evolve a closed expression that calculates the BCS from them. The results show that there are several machine learning classifiers that could be successfully used for body condition scoring. The symbolic regression experiments led to a compact function that only uses two angles for calculating the BCS with an error similar to the error between experts. This function was chosen over the machine learning classifiers since it can be easily transferred to a computer program and makes use of well recognisable features.

Finally, in order to combine the ASMs and the function that estimates the BCS with the objective of building the complete system, the parameters of five ASMs with different local appearance models were tuned using an evolutionary algorithm. Their performance was calculated by applying this function to the shapes obtained from each ASM. The results are quite promising as a reasonable error is obtained for the system, although they are still not completely conclusive. Most of the problem lies in the difficulties of the ASMs when dealing with some of the images, which present quite poor quality due to the fact that they are taken from hand held cameras in uncontrolled, poorly illuminated and very cluttered environments in farms. Consequently, future work will involve the development of an automatic procedure to detect inadequate images when taken and advising the taker to improve the image capture process.

References

1. Bewley, J.M., Schutz, M.M.: An interdisciplinary review of body condition scoring for dairy cattle. Prof. Anim. Sci. **24**, 507–529 (2008)
2. Cootes, T.F., Taylor, C.J.: Active shape models—'smart snakes'. In: Proceedings of the British Machine Vision Conference, Springer-Verlag, pp. 266–275 (1992)
3. Upham, G.: The use of body condition scores in grouping of lactating cows. Compend. Continuing Educ. Pract. Vet. **12**, 581–589 (1990)

4. Leroy, T., Aerts, J.M., Eeman, J., Maltz, E., Stojanovski, G., Berckmans, D.: Automatic Determination of Body Condition Score of Cows Based on 2D Images. Wageningen Academic Publishers, Netherlands (2005)
5. Coffey, M.P., Simm, G., Hill, W.G., Brotherstone, S.: Genetic evaluations of dairy bulls for daughter energy balance profiles using linear type scores and body condition score analyzed using random regression. J. Dairy Sci. **86**, 2205–2212 (2003)
6. Ferguson, J.D., Azzaro, G., Licitra, G.: Body condition assessment using digital images. J. Dairy Sci. **89**, 3833–3841 (2006)
7. Bewley, J.M., Peacock, A.M., Lewis, O., Boyce, R.E., Roberts, D.J., Coffey, M.P., Kenyon, S.J., Schutz, M.M.: Potential for estimation of body condition scores in dairy cattle from digital images. J. Dairy Sci. **91**, 3439–3453 (2008)
8. Halachmi, I., Polak, P., Roberts, D.J., Klopcic, M.: Cow body shape and automation of condition scoring. J. Dairy Sci. **91**(11), 4444–4451 (2008)
9. Azzaro, G., Caccamo, M., Ferguson, J., Battiato, S., Farinella, G., Guarnera, G., Puglisi, G., Petriglieri, R., Licitra, G.: Objective estimation of body condition score by modeling cow body shape from digital images. J. Dairy Sci. **94**(4), 2126–2137 (2011)
10. Cootes, T.F., Taylor, C.J.: Using grey-level models to improve active shape model search. In: Pattern Recognition, 1994. Vol. 1—Conference A: Computer Vision & Image Processing., Proceedings of the 12th IAPR International Conference on. vol 1. pp. 63–67 (1994)
11. Tedín, R., Becerra, J.A., Duro, R.J.: Using classifiers as heuristics to describe local structure in active shape models with small training sets. Pattern Recogn. Lett. 34, 1710–1718 (2013). doi:10.1016/j.patrec.2013.04.026
12. Wang, Y., Witten, I.H.: Inducing model trees for continuous classes. In: Proceedings of the 9th European Conference on Machine Learning Poster Papers, pp. 128–137 (1997)
13. Hall, M., Frank, E., Holmes, G., Pfahringer, B., Reutemann, P., Witten, I.H.: The WEKA data mining software: an update. SIGKDD Explor. Newsl. **11**(1), 10–18 (2009)
14. Koschan, A., Kang, S., Paik, J., Abidi, B., Abidi, M.: Color active shape models for tracking non-rigid objects. Pattern Recogn. Lett. **24**(11), 1751–1765 (2003)
15. Viola, P., Jones, M.: Robust real-time object detection. Int. J. Comput. Vision **57**(2), 137–154 (2002)
16. Koza, J.R.: Genetic Programming: On the Programming of computers by Means of Natural Selection. MIT Press, Cambridge (1992)
17. Storn, R., Price, K.V.: Differential evolution—a simple and efficient heuristic for global optimization over continuous spaces. J. Global Optim. **11**, 341–359 (1997)
18. Caamaño, P., Bellas, F., Becerra, J.A., Duro, R.J.: Application domain study of evolutionary algorithms in optimization problems. In: Proceedings of the GECCO 2008, pp. 377–384 (2008)
19. Caamaño, P., Tedín, R., Paz-Lopez, A., Becerra, J.A.: Jeaf: A java evolutionary algorithm framework. In: IEEE Congress on Evolutionary Computation, IEEE, pp. 1–8 (2010)

Chapter 11
The Impact of Network Characteristics on the Accuracy of Spatial Web Performance Forecasts

Leszek Borzemski and Anna Kamińska-Chuchmała

Abstract Geostatistical methods are very useful tools for spatial forecasts in many research domains. The usage of these methods in the computer science domain is still in its infancy. There were attempts to use simulation and estimation geostatistical methods used in previous research to spatial Web systems performance forecasts. As the results are quite encouraging, therefore, authors decided to carry out further investigations in this field. This chapter presents an overview of the work concerning the use of geostatistical methods and the analysis of the impact of various factors on the accuracy of spatial Web performance forecasts for servers belonging to different Autonomous Systems (ASs). Forecasts were made by using geostatistical methods. The data for research was collected in active Internet Web Performance measurements carried out by software agents monitoring a group of Web servers. In this research, the network routes from agents in Gdańsk and Wrocław to European Web servers are considered.

Keywords Forecast · Geostatistics · Internet · IoT · Performance · MWING · WoT · Web

11.1 Introduction

The plan to develop the Internet of Things (IoT) [1, 2], as well as, the Web of Things (WoT) [3, 4] becomes to be more realistic. Nowadays, mobile things (devices), for example tablets, smartphones or smartbooks are very popular and used especially by business and young people. Radio-frequency identification (RFID) has a wide

L. Borzemski (✉) · A. Kamińska-Chuchmała
Institute of Informatics, Wrocław Uniwersity of Technology, 50-370 Wrocław, Poland
e-mail: leszek.borzemski@pwr.wroc.pl

A. Kamińska-Chuchmała
e-mail: anna.kaminska-chuchmala@pwr.wroc.pl

J. W. Tweedale and L. C. Jain (eds.), *Recent Advances in Knowledge-based Paradigms and Applications*, Advances in Intelligent Systems and Computing 234,
DOI: 10.1007/978-3-319-01649-8_11, © Springer International Publishing Switzerland 2014

usage in many industries, retail, access control, smart-cards and this is only one step to IoT. All mobile devices are based on wireless or adhoc communications. People, whoever they are, want to have access to Internet and Web by their mobile things. Therefore, good quality of Web performance is to be addressed. Assuming that Internet is everywhere, one can consider Web server and connections between them in a given geographical space, that perfectly fits to using geostatistical methods.

In Sect. 11.2 the related work with using geostatistical methods is described. In Sect. 11.3, the database and results of our previous research are shortly presented. Next, in Sect. 11.4 the analysis of the impact of network characteristics on the accuracy of spatial Web performance forecasts for servers belonging to different Autonomous Systems (ASs) will be presented. To answer, if the geographical distance and network distance measured by the Round-Trip Time (RTT) have influence on forecast accuracy. Conclusions and future research are presented in Sect. 11.5.

11.2 Related Work and Background of Associated Technology

The term geostatistics was created in the early 1960s, when French engineer Georges Matheron formalized the theory of this approach [5]. The geostatistics refers to spatial aspects due to the prefix"geo" [6]. The geostatistical methods originate from the Kriging - the estimation method developed by Daniel Krige [7].

The largest research center and the cradle of knowledge bringing together the finest specialists in geostatisticst is in the Center for Geostatistics of the Ecole des Mines de Paris in Fontainebleau.[1]

Till late-1990s geostatistics focused on geology, especially in solving the problems in petroleum drilling, such as realistic heterogeneity models for unbiased flow predictions [8, 9]. The volume of rock sample typically represents only a minute fraction of the total volume of a hydrocarbon reservoir. Even in such cases, when variables are sampled very sparsely the geostatistical methods can deal with the problem of estimation of studied variable in given area.

Later geostatistics has been applied to problems of spatial modeling and uncertainty in enviromental studies (air quality monitoring), hydrogeology, and agriculture. For example, it issued is used to study of risk related to air pollution [10]. The diffusion of a pollutant in a geographical space is driven by complex physical and chemical laws influenced by local environmental parameters, it spreads out in a three dimensional space. The authors proposed a probabilistic solution, which consists of specializing concentrations obtained from an air quality monitoring network. They use two geostatistics models: non linear estimation methods and simulation methods showing annual mean concentration in NO_2 in the city of Rouen, France. Another research in the field of oceanography use geostatistical simulation method Turning Bands to model marine ecosystem [11]. The results based on two different datasets

[1] See http://www.geosciences.mines-paristech.fr/en

where compared, and simulation models for distribution of three different variables: phytoplankton, nitrate, and temperature where created. Problems related to agriculture are considered especially for poor regions of developing countries. For example they are addressed in [12], where the author has applied the methods in a precise agriculture framework. Oliver's research is likely to have more impact as it becomes increasingly possible to obtain data cheaply in conjuction with more farmers using onboard digital maps of soil and crops to manage their production.

Nowadays the usage of geostatistical methods includes also such fields as forestry, biomass estimation, nuclear decommissioning and decontamination, epidemiology, geochemistry, meteorology. The main aim of Tang and Hossain [13] is to assess spatial interpolation schemes for transfer of the error characteristics from ground validation regions to non-ground validation regions. They use kriging methods (ordinary, indicator and disjunctive) for spatial transfer of error metrics.

There also appeared new branches of science where geostatistical methods are useful. Quite good example is economic analysis [14], where authors introduce a new way of investigating linear and nonlinear Granger causality between exports, imports and economic growth in France over the period 1961–2006 with using geostatistical models: kiriging and inverse distance weighting.

One of the authors used geostatistical simulation methods: Turning Bands (TB) and Sequential Gaussian Simulation (SGS) to spatial electric loads forecasting [15, 16]. Geostatistical simulation methods were applied to spatial electric load forecasting distribution and transmission networks for area of Poland with forward period of time equal to one year.

Geostatistical methods were also used to study the problem of spatial distribution of floating car speed analyzed by exact floating car speed data of the study area in Beijing [17], that seems to be similar to the problem of traffic data packets on the Internet. Currently, to the best of the authors knowledge the geospatial approach to Web performance prediction presented in this chapter is unique as developed in our chapters, leaving no similar problem statement in the literature.

The experiment generates spatial Web server performance forecasts with using geostatistical simulation methods: TB [18, 19] and SGS [20] and estimation method—Simple Kriging (SK) [21]. This novel research was made on the database collected Multi-agent Web pING (MWING) [22, 23] active experiments, which will be thoroughly presented in the Sect. 11.3.

11.3 Experimental Setup and Spatial Models of Forecasts

Databases which were used in forecasts, collect performance data gathered during active measurements made by the Multiagent Internet Measurement system called MWING, which is the global Internet measurement infrastructure developed in our Institute [22, 23] where agents were installed on local hosts in networks belonging to academic campuses in four geographical locations: Gdańsk, Wrocław, Gliwice (in Poland) and Las Vegas (in the USA). Every agent monitored regularly over 60 Web servers localized worldwide.

One of the databases was collected by MWING is agent located in Gdańsk, target by means of HyperText Transfer Protocol (HTTP) transactions several European Web servers. The database contained the information about a server's geographical location, which the Gdańsk agent targeted, web perfomance index-Z, which was the total downloading time of rfc1945.txt file and the timestamp of taking a measurement. The database collected measurements from 7th to 28th of February 2009 and they were taken every day at the same time: at 06:00 a.m., 12:00 a.m., and 6:00 p.m.

The next database was collected by MWING agent located in Wrocław and has collected the period between 1st and 28th February 2009. The measurements were taken once a week on Monday at fixed times: 06:00 a.m., 12:00 a.m., and 6:00 p.m.

The spatio-temporal forecasts of Web servers performance based on database collected by agent from Gdańsk was calculated with using: TB [24], SGS [20] and SK [21] methods, respectively. SGS is the method based on Bayes' theorem and Monte Carlo simulation. It allows for Z_s multidimensional realizations of Gaussian random function [25].

The TB method created by Matheron is a stereologic tool used for the reduction of multidimensional simulation to a one-dimensional one [26]. The idea of the TB method is the reduction of random simulation of a Gaussian function with covariance C to the simulation of an independent stochastic process with covariance C_θ.

SK method is a spatial regression named also as kriging with known mean. SK is used to estimate residuals, where average m is given a priori. Kriging is preceded by an analysis of the spatial structure of the data. The representation of the average spatial variability is integrated into the estimation procedure in the form of a variogram model. For more information see Wackernagel [27]. Three forecasting models for 6:00 a.m., 12:00 a.m. and 6:00 p.m. in each method were created.

Borzemski and Kaminska-Chuchmała [21] presented the comparison of three geostatistical methods. In Table 11.1 it could be seen that errors of forecast SGS method are the smallest, from 1 to 4 % less than in other methods. Therefore, simulation methods are better than estimation ones, which give worse results especially in cases when Z distribution has large skewness. It indicates that simulation method further introduces a changeability of examined process. The measurement error caused by, for example, improper selection of variogram model gives only one exception where TB method obtains larger error of forecast than SK (the difference is about 1.5 %).

Table 11.1 Geostatistical method mean forecasted error *ext post* for all Web servers in a four-day forecast [21]

Mean forecasted error	6:00 a.m.	12:00 a.m.	06:00 p.m.
For SGS	24.83 %	16.06 %	18.53 %
For TB	26.91 %	20.00 %	17.55 %
For SK	30.87 %	18.47 %	19.11 %

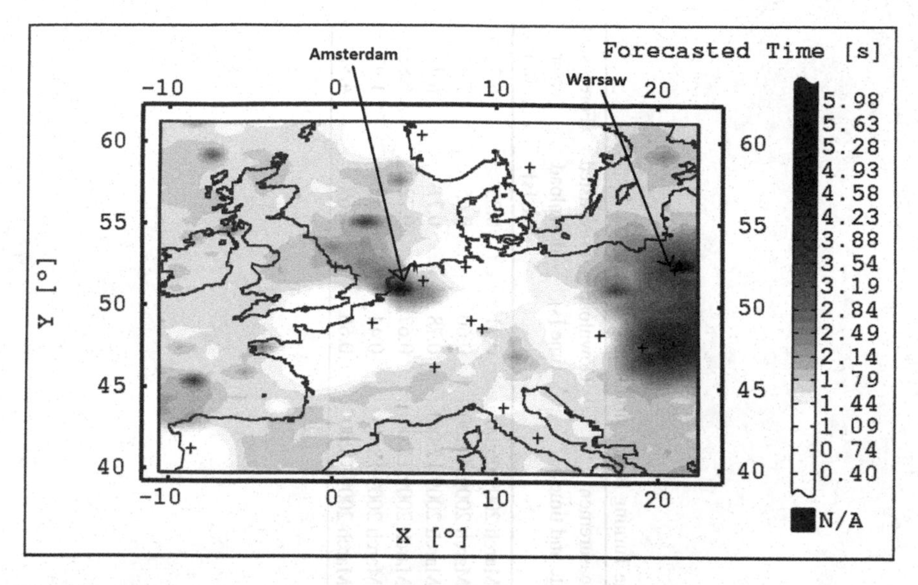

Fig. 11.1 Sample raster map of Web download time values from the Internet on 01/03/2009 at 06:00 p.m. [20]

The final effects of the forecast are presented in Fig. 11.1. This raster map for the 1st day of prognosis (01/03/2009) at 06:00 p.m. presents the download time from the European Web servers, where each cross corresponds to a different Web server. This figure describes how varied is Web performance due to different research hours and days, because geostatistical methods could give information about performance not only for considered servers, but for a whole considered area. However, in Fig. 11.1 there are two servers with the largest download time located in Amsterdam, Netherlands and Warsaw, Poland.

The forecast model created on the basis of the database collected by agent from Wrocław, presented by authors in [28], was used to predict the total time of resource download from the Internet depending on the forecasted hour on Monday. Spatial forecast was calculated with a two-week time advance. It encompassed the period between 1st and 14th March 2009. Table 11.2 presents a few exemplary Web servers in different hours and the average, absolute percentage, relative forecast error *ex post* calculated for them. Analyzing the results presented in Table 11.2, one can observe that the forecast prepared using the TB method represents the actual download time quite well. Unfortunately due to the big span of Web servers in the examined area and high measurement data dispersion, not all of the results were characterized with good accuracy of prediction results.

Table 11.2 Exemplary results of spatial forecast of Web servers performance calculated with the Turning Bands simulation method [28]

URL address	Country/city	Measurement date and time	Download time [s]	Forecasted download time [s]	Forecasted error *ex post* [%]
http://wigwam.sztaki.hu/rfc/rfc1945.txt	Hungary/Budapest	2 March 2009,06:00	0.31	0.33	6.71
http://omega.di.unipi.it/local/home/rfc/rfc1945.txt	Italy/Pisa	9 March 2009, 06:00	0.60	0.65	7.86
http://www-uxsup.csx.cam.ac.uk/pub/doc/rfc/rfc1945.txt	United Kingdom /Cambridge	2 March 2009,12:00	0.78	0.71	8.42
http://ftp.univie.ac.at/netinfo/rfc/rfc1945.txt	Austria/Vienna	9 March 2009,12:00	0.67	0.72	6.88
http://curl.nedmirror.nl/rfc/rfc1945.txt	Netherlands/Eindhoven	2 March 2009,18:00	0.44	0.43	1.97
http://paginas.fe.up.pt/~jvv/net/rfc1945.txt	Portugal/Porto	9 March 2009,18:00	0.66	0.69	4.82

11.4 The Impact of Network Characteristics on the Accuracy of Spatial Web Performance Forecasts for Servers Belonging to Different Autonomous Systems

Having regard two databases described in previous section, the impact of various factors on the accuracy of spatial Web Performance forecasts was determined. In this analysis we consider servers belonging to different ASs. The databeses includes 17 Web servers that belong to different ASs with only one excepction. Autonomous Systems Number (ASN) was checked based on the Classless Inter-Domain Routing (CIDR) Report.[2] Routes from Wrocław or Gdańsk in Poland target to Web server targeted were check by *traceroute* command. We measured the network distance between ASs by the average Round Trip Time (RTT). The Internet traffic coming out from Poland via AS8501 officially located in Poznań. In the next step, the traffic was supported by one of two ASs: AS20965 in Cambridge, UK (illustrated in Fig. 11.2) or AS2603 in Stockholm, SE (Fig. 11.3).

After analysis one can report the following remarks:

- There are two main ASs which support the traffic from Poland to Europe. One could clearly see when regard the Figs. 11.2 and 11.3 that AS20965 from Cambridge have twice more Round Trip Time (RTT) then AS2603 on average.

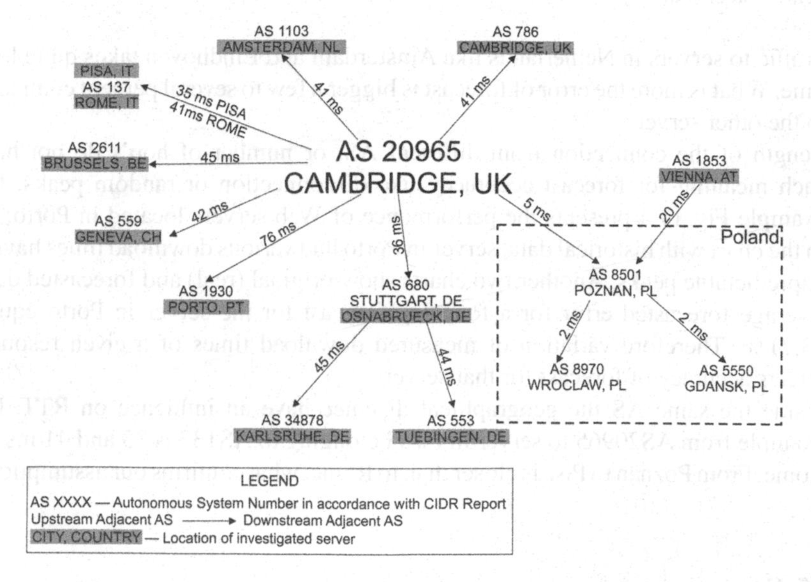

Fig. 11.2 Graph of connections between servers,including servers belonging to the Autonomous Systems–Cambridge

[2] See http://www.cidr-report.org

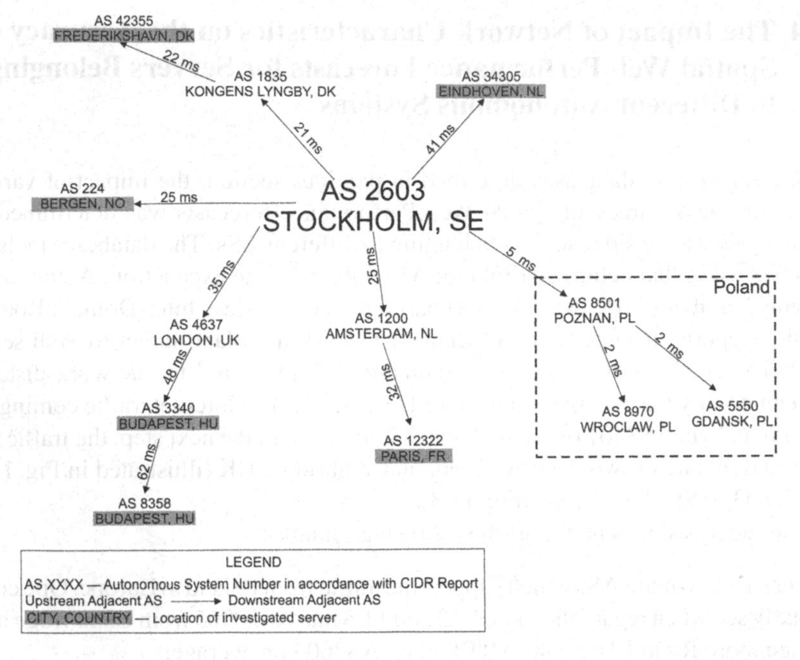

Fig. 11.3 Graph of connections between servers, including servers belonging to the Autonomous Systems – Stockholm

- Traffic to servers in Netherlands like Amsterdam and Eindhoven takes quite long time. What is more the error of forecast is bigger a few to several percent compared to the other servers.
- Length of the connection from different AS, or number of hop's do not have such meaning for forecast accuraccy like disconnection or random peaks. For example Fig. 11.4 presents the performance of Web servers located in Porto, Pt. In the chart with historical data, server in Porto had various download times having unpredictable peaks. Another two charts show original (real) and forecasted data. Average forecasted error for a four-day forecast for the server in Porto equals 23.71 %. Therefore variation of measured download times of a given resource affects accuracy of forecast for that server.
- Inside the same AS the geographical distance have an influence on RTT. For example from AS20965 to server in Pisa belonging to AS137 is 35 and 41 ms for Rome. From Poznań to Pisa is closer than to Rome, what confirms our assumptions.

11.5 Summary

This chapter presents Web performance forecasting using the TB, SGS and SK geostatistical methods, which is an innovative approach in considered research area. Large-scale measurement experimenting MWING system was performed in a

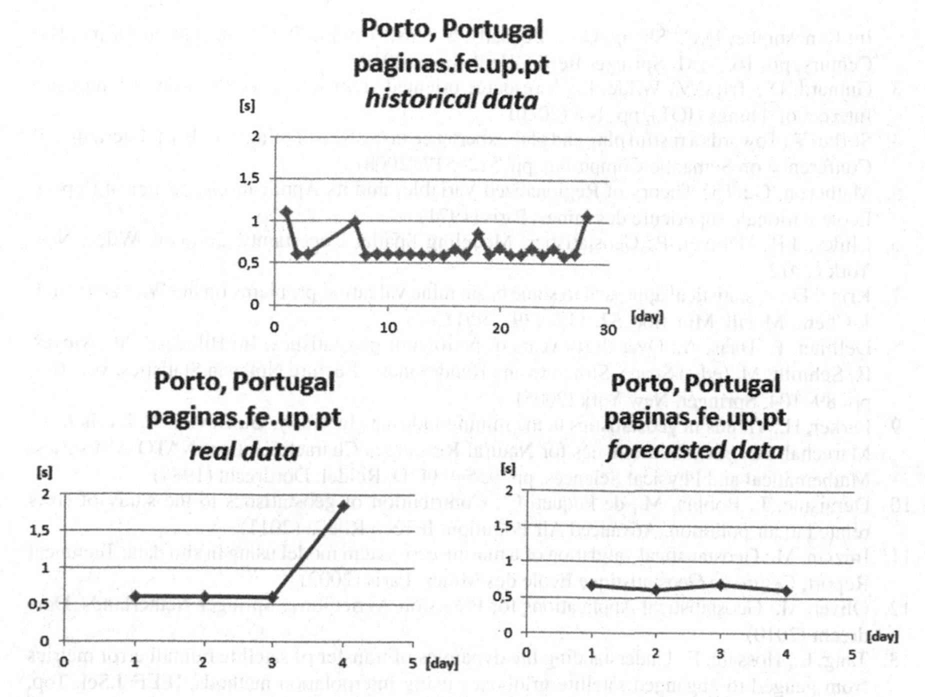

Fig. 11.4 Comparison of download times from the Web server in Porto for 6:00 a.m. [20]

real-life Internet to gather the data characterizing performance of many Web servers localized in Europe and perceived from agent installed in Gdańsk and Wrocław. The impact of various factors on the accuracy of spatial Web performance forecasts using geostatistical methods for servers belonging to different Autonomous Systems was analyzed. As conclusion we can claim, that there is relationship between forecasts accuracy and fluent traffic on routes to servers. Geographical distance has an influence on accuracy of forecast but network distance has not.

There is many factors which could have influence and could improve or decrease forecast accuracy. Further investigation are planned in the future. The authors are aware that our conclusions are yet tentative and they should be confirmed many times in other real-life experiments. Other network-related factors that may have influence on Web performance forecasts should be consider.

References

1. Internet of Things - An Action Plan for Europe. Commission of the European Communities, Brussels, COM (2009) 278
2. Uckelmann, D., Isenberg, M.A., Teucke, M., Halfar, H., Scholz-Reiter, B.: Autonomous control and the internet of things: Increasing robustness, scalability and agility in logistic networks.

In: Ranasinghe, D.C., Sheng, Q.C., Zeadally, S. (eds.) Unique Radio Innovation for the 21st Century, pp. 163–181. Springer Berlin, Heidelberg (2010)
3. Guinard, D., Trifa, V., Wilde, E.: A resource oriented architecture for the web of things. In: Internet of Things (IOT), pp. 1–8 (2010)
4. Stirbu, V.: Towards a restful plug and play experience in the web of things. In: IEEE International Conference on Semantic Computing, pp. 512–517 (2008)
5. Matheron, G.: The Theory of Regionalized Variables and its Applications. Technical Report. Ecole nationale superieure des mines, Paris (1971)
6. Chiles, J.P., Delfiner, P.: Geostatistics: Modeling Spatial Uncertainty, 2nd edn. Wiley, New York (2012)
7. Krige, D.: A statistical approach to some basic mine valuation problems on the Witwatersrand. J. Chem. Metall. Min. Soc. **52**, 119–139 (1951)
8. Delfiner, P., Haas, A.: Over thirty years of petroleum geostatistics. In: Bilodeau, M., Meyer, F., Schmitt, M. (eds.) Space, Structure and Randomness, Lecture Notes in Statistics, vol. 183, pp. 89–104. Springer, New York (2005)
9. Parker, H.: Trends in geostatistics in the mining industry. In: Verly, G., David, D., Journel, A., Marechal, A. (eds.) Geostatistics for Natural Resources Characterization, NATO ASI series: Mathematical and Physical Sciences, pp. 915–934. D. Reidel, Dordrecht (1984)
10. Deraisme, J., Bobbia, M., de Foquet, C.: Contribution of geostatistics to the study of risks related to air pollution. Advanced Air Pollution, InTech Rijeka (2011)
11. Inizian, M.: Geostatistical validation of a marine ecosystem model using in situ data. Technical Report. Centre de Geostatistique Ecole des Mines, Paris (2002)
12. Oliver, M.: Geostatistical Applications for Precision Agriculture. Springer Netherlands, Dordrecht (2010)
13. Tang, L., Hossain, F.: Understanding the dynamics of transfer of satellite rainfall error metrics from gauged to ungauged satellite gridboxes using interpolation methods. IEEE J.Sel. Top. Appl. Earth Observ. Remote Sens. **4**(4), 844–856 (2011)
14. Amiri, A., Gerdtham, U.: Relationship between exports, imports, and economic growth in france: evidence from cointegration analysis and granger causality with using geostatistical models. munich personal repec archive chapter no. 34190 (2011)
15. Kamińska-Chuchmała, A., Wilczyński, A.: Application simulation methods to spatial electric load forecasting. Rynek Energii **80**(1), 2–9 (2009) (in Polish)
16. Kamińska-Chuchmała, A., Wilczyński, A.: Spatial electric load forecasting in transmission networks with sequential gaussian simulation method. Rynek Energii **92**(1), 35–40 (2011)
17. Wang, Y., Zhuang, D., Liu, H.: Spatial distribution of floating car speed. J.Transp. Syst. Eng. Inf. Technol. **12**(1), 36–41 (2012)
18. Borzemski, L., Kamińska-Chuchmała, A.: Client-perceived web performance knowledge discovery through turning bands method. Cybern. Syst. **43**(4), 354–368 (2012)
19. Borzemski, L., Kamińska-Chuchmała, A.: Distributed web systems performance forecasting using turning bands method. IEEE Trans. Industr. Inf. **9**(1), 254–261 (2013)
20. Borzemski, L., Kamińska-Chuchmała, A.: Knowledge engineering relating to spatial web performance forecasting with sequential gaussian simulation method. In: Graña, M., Toro, C., Posada, J., Howlett, R., Jain, L.C. (eds.) Advances in Knowledge-Based and Intelligent Information and Engineering Systems, Frontiers in Artificial Intelligence and Applications, vol. 243, pp. 1439–1448. IOS Press, Amsterdam (2012)
21. Borzemski, L., Kamińska-Chuchmała, A.: Web performance forecasting with kriging method. In: Ali, M., Bosse, T., Hindriks, K.V., Hoogendoorn, M., Jonker, C.M., Treur, J. (eds.) Contemporary Challenges and Solutions in Applied Artificial Intelligence, Studies in Computational Intelligence. vol. 489, pp. 149–154, Springer-Verlag (2013)
22. Borzemski, L.: The experimental design for data mining to discover web performance issues in a wide area network. Cybern. Syst. **41**(1), 31–45 (2010)
23. Borzemski, L., Cichocki, L., Fraś, M., Kliber, M., Nowak, Z.: MWING: A multiagent system for web site measurements. In: Nguyen, N.T., Grzech, A., Howlett, R.J., Jain, L.C. (eds.) Agent and Multi-Agent Systems: Technologies and Applications, Lecture Notes in Computer Science, vol. 4496, pp. 278–287. Springer Berlin Heidelberg (2007)

24. Borzemski, L., Danielak, M., Kamińska-Chuchmała, A.: Short-term spatio-temporal forecasts of web performance by means of turning bands method. In: Nguyen, N.T., Hoang, K., Jedrzejowicz, P. (eds.) Computational Collective Intelligence. Technologies and Applications, Lecture Notes in Computer Science, vol. 7654, pp. 132–141. Springer Berlin Heidelberg (2012)
25. Leuangthong, O., Khan, K., Deutsch, C.: Solved Problems in Geostatistics. Wiley, New Jersey (2008)
26. Lantuejoul, C.: Geostatistical Simulation: Models and Algorithms. Springer Berlin, Heidelberg (2002)
27. Wackernagel, H.: Multivariate Geostatistics. Springer Berlin, Heidelberg (2003)
28. Borzemski, L., Kamińska-Chuchmała, A.: Knowledge discovery about web performance with geostatistical turning bands method. In: Koenig, A., Dengel, A., Hinkelmann, K., Kise, K., Howlett, R.J., Jain, L.C. (eds.) Knowlege-Based and Intelligent Information and Engineering Systems, Lecture Notes in Computer Science, vol. 6882, pp. 581–590. Springer Berlin Heidelberg (2011)

Chapter 12
Using Multi-Agent Systems Technique for Developing an Autonomous Model Used to Analyze Work-Stress Data

Anusua Ghosh, Jeffery W. Tweedale and Andrew Nafalski

Abstract This chapter presents a semi-autonomous data collection and analysis application that facilitates the measurement of indivudal stress (or any other metric) within the workplace. This novel approach uses a hybridized autonomous Multi-Agent System (MAS) framework that has been tailored to analyze work-related stress for individuals submitting the on-line survey in real-time. Psychologists continue to report that work-stress affects people from all profession. This model allows them to remotely measure the level of stress within the workplace, against national norms, that can help employers identify and address or prevent stress by influencing changes to the working environment. The system has been presented at conferences, then tested both within the workplace and remotely as an on-line kiosk. Intelligent Multi-Agent Decision Analyser (IMADA) is the core component of the model and each agent represents independent capabilities in their own right. The MAS provides the interaction and communication between each agent in order to accomplish the desired desired goal (individual bench-marked responses). IMADA replaces the existing manual processes and reduces the significant effort required to access employees. It also automates the collection, analysis and management of the survey data (which was typically achieved by professionals through questions during long phone interviews).

Keywords Artificial Intelligence · Intelligent Agent · Multi-Agent System · Neural Network · Fuzzy Logic

A. Ghosh (✉) · J. W. Tweedale · A. Nafalski
School of Electrical and Information Engineering, University of South Australia, Adelaide, Australia
e-mail: ghosh@unisa.edu.au

A. Nafalski
e-mail: andrew.nafalski@unisa.edu.au

J. W. Tweedale
Aerospace Division, Defence Science and Technology Organization, Edinburgh, Adelaide, Australia
e-mail: jeffrey.tweedale@unisa.edu.au; jeffrey.tweedale@dsto.defence.gov.au

J. W. Tweedale and L. C. Jain (eds.), *Recent Advances in Knowledge-based Paradigms and Applications*, Advances in Intelligent Systems and Computing 234, DOI: 10.1007/978-3-319-01649-8_12,

12.1 Introduction

The field of computer science provides a significant suite of tools that researchers can use to solve problems. Heuristics techniques have been used for decades to solve computational problems. Artificial Intelligence (AI) and Multi-Agent System (MAS) techniques have been more recently used to solve real-world issues. As society migrates into a digital future, data collection problems continue to challenge everyone. The volume of collected data is increasing exponentially, requiring more autonomous techniques to assist non-computer specialist to analyse there own collections of information. This process can be complex, forcing researchers to develop tools that seamlessly integrate the collection and analysis mechanisms that derive usable information in real-time. This chapter presents the design, development and implementation of a hybridized autonomous tools that was developed using MAS techniques to collection and analysis information on behalf of qualified psychologists. The original research question was aimed at "*analyzing work-stress data as the user participates in an on-line survey and submits personalised responses. They would receive feedback compared against normalised benchmarks in real-time*". Conceptually the approach chosen is shown in Fig. 12.1.

This model describes the evolution from the existing manual process (using using a Computer Aided Telephone Interviews (CATIs), through to the creation of a semi-automated survey collection system called the Intelligent Agent Framework (IAF). The collected data was pre-processed into spreadsheets, based on the need for psychological analysis and then statistically analyzed using Statistical Package for Social Sciences (SPSS). The results of the analyzed data was originally reported by Dollard and Taylor [1]. The IAF would eventually be transformed into an autonomous on-line application called Intelligent Multi-Agent Decision Analyser (IMADA). The original goal was to enable the homogenous collection of employee stress data, using on-line systems, that could be access through the internet or kiosk style workstaitons.

The developed model categorizes and benchmarks nationally users stress level, providing feedback in the form of graph in real-time. The model was implemented and tested using the e-portal StressCafé, allowing user of the system to submit an on-line survey in relation to their work-stress. The results demonstrated that using this model to measure work-stress levels, can provide a first step towards identification, prevention, and providing smart changes to the working environment.

This chapter describes that evolution. Section 12.2 describes the background knowledge required. This is followed by the design and methodology of the

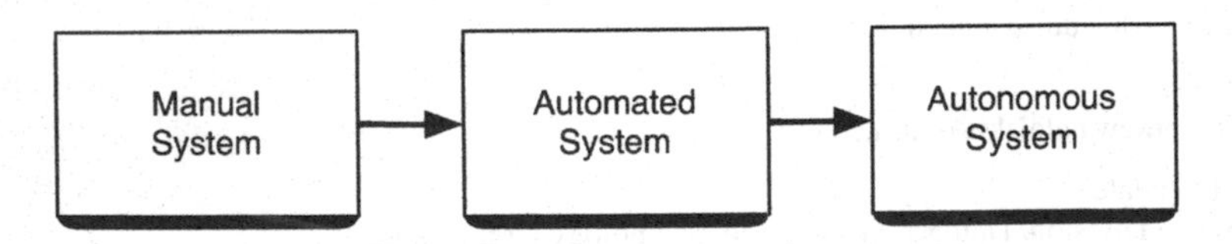

Fig. 12.1 Transformation of the model from manual to autonomous

automated model in Sect. 12.3. An outline of the development behind the autonomous model using hybridized MAS techniques is provided in Sect. 12.4. A case study is discussed in Sect. 12.5 followed by the conclusion and future effort.

12.2 Background Knowledge

Computers use binary logic in order to store and manipulate data. methods of capturing and processing human-like information in digital form, have been evolved over time. The goal of making computers do human-like things surfaced early in this evolution. In 1956, the Dartmouth conference hosted a group of experts as a think-tank to champion this topic. During the conference, John McCarty introduced the term AI. He defined this term as the science and engineering required to make intelligent machines. This group understood that Human beings are born with some form of intelligence that evolves over time. This helped them understand, how to develop and create machines to do human-like activities. Machines, on the other hand are built by humans, who need to embody functionality to enable them to appear to behave intelligently and operate independently. The science behind Machine Intelligence (MI) has focused on using AI techniques to achieve this functionality. The field has been defined in terms of and system designs now employ Intelligent Agent (IA) [2, 3]. For this chapter, the term intelligent and intelligence is stated as the computational functions representing the ability to make decisons in order to achieve goals in the real world. An agent is a system that perceives its environment and then takes appropriate action to gain success. Nilsson provides an insight via a comprehensive history of AI [4]. Modern applications use IAs within computer programs to emulate aspects of intelligence that act independent of their users [5, 6]. An agent can manifest in software and hardware (firmware) as entities that autonomously react to changes in the environment, using sensors and actuators [7]. Whilst there is some debate regarding the exact definition of the term IA, most researchers accept a widely used definition by Wooldridge, that considers an intelligent agent to be both flexible and capable of operating autonomously to meet its design objectives [8]. To further define the concept, IAs are also considered to be reactive and capable of adapting in real time to accommodate any changes in the environment [7]. In order to manage the complexity of developing software, system designers and software engineers use high level abstraction when developing applications. This enables them to focus on the important and essential properties of a problem rather than the incidental component of that problem.

Russell believes that agents are most commonly used in challenging or changing, dynamic, unpredictable, and unreliable environments [9]. An autonomous agent is capable of accomplishing goals without the direct intervention by humans or other agents, maintaining control over its own actions and internal state [8]. The Amazon website extensively employs intelligent agent technology to develop its online book store. Other examples are discussed in Tweedale et al. [10]. Modern solutions employ mutliple techniques and IA capabiilities are often teamed to create more complex

MASs. IAs are capable of learning quickly, using large volumes of data, and adapt in real time. Examples include the combination of one or more agent in a system; such as the MAS framework supporting the IAF and IMADA. The latter uses MAS, Artificial Neural Network (ANN), Fuzzy Logic (FL) and hybridized systems to deliver survey results in real-time. Each concept is discussed in Sects. 12.2.1–12.2.4.

12.2.1 Multi-Agent System

MAS framework represents a new way of conceptualizing and implementing distributed software because they are composed of multiple interacting computing elements, known as agents [7]. In MAS agents communicate and interact with each other to achieve a common goal. MAS technology can be used to integrate a legacy data capabilities tool to enhance diagnostic support to engineers in software applications [11]. This technology also enables designers to explore the sociological and psychological foundations as they deal with coordinating intelligent behavior to accomplish common goals [12]. The MAS domain is an active area for researchers. It is being used to develop new applications that operate in the human-computer environment [13]. MAS can be regarded as a rational, real-time interactive agent, working collaboratively to achieve a common goal. In some MAS, agents can communicate with other agent within the team. Although in many other systems, agents only communicate with agents that are directly linked to them [14]. The next section discusses application using IA and MAS.

Intelligent agents are frameworks or architectures that are used to develop software applications. These techniques have been hailed as the breakthrough and revolutionary in software development since the 1990's [7]. Agent, MAS technologies, methods and theories continue to contribute to many diverse applications. Agents are flexible, so can be equipped with capabilities such as learning, reasoning and mobility, which enable agents to autonomously carry out their goals. Using an agent oriented approach, both simple and complex problems can be decomposed into different components (agents) or capabilities. Each agent and interact to achieve a common goal.

An intelligent agent based monitoring platform for application in engineering is discussed in a chapter by Mangina where the approach of combining two or more techniques are considered [15]. Weaknesses can be reduced by combining or integrating different techniques of different strengths, therefore generating a hybrid solution [15]. Hybrid solutions are created to neutralize weaknesses and optimize application software. In the field of medical science IA and MAS are an adequate tool for tackling health care problems. An example IA based application called Graphical User Interface (GUI) for *e-physician* has been developed to help physicians get online access to patient information. This assists them in prioritizing emergency cases, easy access to laboratory results, which reduces overall time and cost [16]. It has also been shown that the use of an intelligent GUI can improve the effectiveness and efficiency of the Human Computer Interaction (HCI) [16, 17]. Fuzzy-means clustering has also been applied to generate rules [18], for implementing an IA that interacts with humans to assist operators in Medical Diagnostic Systems [19].

12.2.2 Artificial Neural Network

The curiosity of understanding of the human brain led neuro-physiologist, Warren McCulloch and mathematician Walter Pitts to write a chapter on how biological neuron could expect to operate. They built a working model in 1943. This contained a number of simple electric circuit [20]. Following this success, Frank Rosenbalt developed the perceptron. In 1958 he delivered the first practical ANN [21]. This ANN was developed based on the biological nervous systems and observations about how the brain works. They are adaptive information processing systems, which consist of highly interconnected processing elements working together to solve problems. A Neural Network is a combination of a digital representation of a human inspired neuron, where the network is the means of creating the virtual brain. A technical neural network consists of simple processing units, the neurons, and directed, weighted connections between those neurons. The function of the neural network is to produce a pattern on its output which is correct for the class when an input is presented to the network. A neural network is a structure that receives inputs process the data, and provides an output, Once an input is presented to the neural network, and a corresponding desired or target response is set at the output, an error is composed from the difference of the desired response and the real system output [22]. ANN are capable of transforming an input vector from n-dimensional space to an output vector in m-dimensional space [23].

Neural networks can be single layer or multi-layered networks. In a single layered network, there is an input layer and an output layer along with weighted edges. The weights are summed; depending on the activation function, an output from the network is generated using a non linear transfer function. The transfer function can be sigmoidal function or a threshold function. In the case of sigmoidal function the input changes gradually where for the latter case the input is processed only when a certain threshold is reached, then switches off again [24]. A perceptron can solve logical functions that are linearly separable. This meant they can be separated into regions using a single line. But all logical function are not linearly separable and cannot be solved using a single perceptron, therefore it may involve the use of more than one perceptron with a feed-forward network (typicalled labelled a multi-layered feed-forward network).

Multi-layered feed-forward network on the other hand consist of an input layer, a hidden layer/layers and an output layer. This network is most commonly used for classification and prediction. The back-propagation algorithm developed independently by Werber, Parker and Rumelhart can be efficiently used to train a multi-layered feed-forward network [25–27]. There are other type of network and algorithm see [23, 24], this research uses a multi-layered feed-forward neural network utilizing the back-propagation algorithm with sigmoid function given in Eq. 12.1.

12.2.3 Fuzzy Logic

Using Fuzzy Logic, machines are able to cope with unknown, unstructured and dynamically changing environments. New systems evolved that are capable of fuzzy approximation, recognition and behavior modeling using human-like terminology.[1] Recent examples include improve position fixing while navigating and improved image segmentation while processing hyper-spectral scenes [28]. Developers have traditionally relied on AI enhanced *Rule-based* controller concepts to explain what should be occurring when solving problems. Zadeh is known as the father of fuzzy logic used logic to represent data as words. The data is categorized using a partial set membership in lieu of the crisp set membership for digital data [29]. With the introduction of membership functions, fuzzy sets can be differentiated. Fuzzy logic can be applied to interpret the output from a neural network, as well as giving a precise interpretation of the data in linguistic terms [24]. This fuzzified reason-based controller uses similar rules to manage uncertainty and imprecision and vice versa. This concept was successfully implemented to control many complex systems and as a means of providing human readable results in IMADA.

12.2.4 Hybrid Intelligent System

Hybrid intelligent systems are computational system that integrates different computational technique; which can then be used to support problem solving and decision making [30, 31]. As different intelligent technique have their advantages and disadvantages, cannot be universally applied to solve any problem, hence integrating the individual intelligent technique and modeling as hybrid system, the limitations of each individual technique is minimized. These systems have multiple parts which interact with each other to solve problem, thus modeling hybrid intelligent system using MAS is suitable. There are three types of hybrid system as [9, 31, 32]:

1. Sequential hybrid system: In Sequential hybrid system, the intelligent component are arranged sequentially, as the output from one component is fed as an input to the other component.
2. Auxiliary hybrid system: Auxiliary hybrid system on the other hand allows one component to call other component as a subroutine to manipulate information accordingly.
3. Embedded hybrid systems: Embedded hybrid system allow the individual component to be fused in such a way that it seems that one cannot perform without the other component.

[1] Examples include a number of simple well defined behaviors; such as: avoidance, reach, follow, align, jump, turn and pass.

12.3 Developing an Autonomous Model

The autonomous model was developed in phases as described in Sects. 12.3.1–12.3.3. Figure 12.2 presents the model and how it has evolved from a manual process to an automated tool and then eventually transformed into an autonomous system.

12.3.1 Manual System

The manual system collected data using survey form via a computer aided telephone interview, where the interviewer respond via telephone survey was recorded using spreadsheet. The data was then pre-processed and analyzed using the software SPSS. The result of the analyses was reported as journal publication, oral presentation in conferences [1].

12.3.2 Automated Model

Automation can be achieved by using machine, information technologies or other control systems that can optimize productivity, increase predictability of quality, improved consistency of process. The IAF was designed to automate the existing manual processes used to collect the CATI. These concepts are outlined in Fig. 12.3.

The framework consists of multiple databases, the graphical user interface, analyser capability (to read data, record individual performance, compare individual performance), knowledge base, learning component and decision processes along with a feedback mechanism. The model processes data automatically, optimizing productivity and predictability.

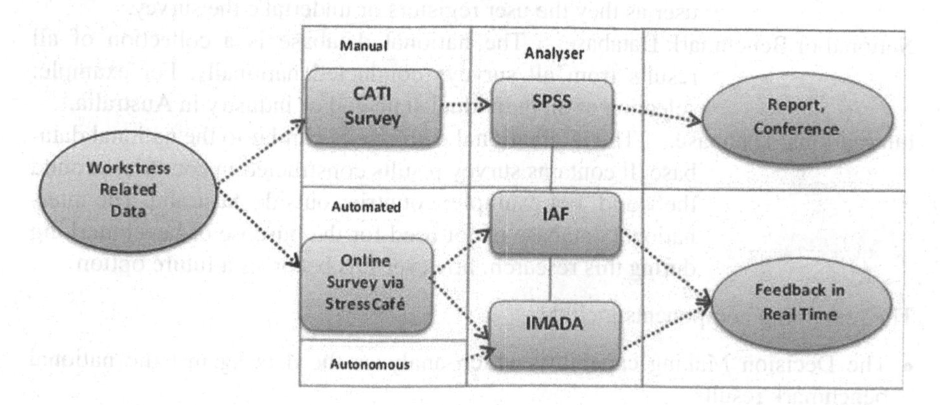

Fig. 12.2 Evolution phases of the model from manual to autonomous

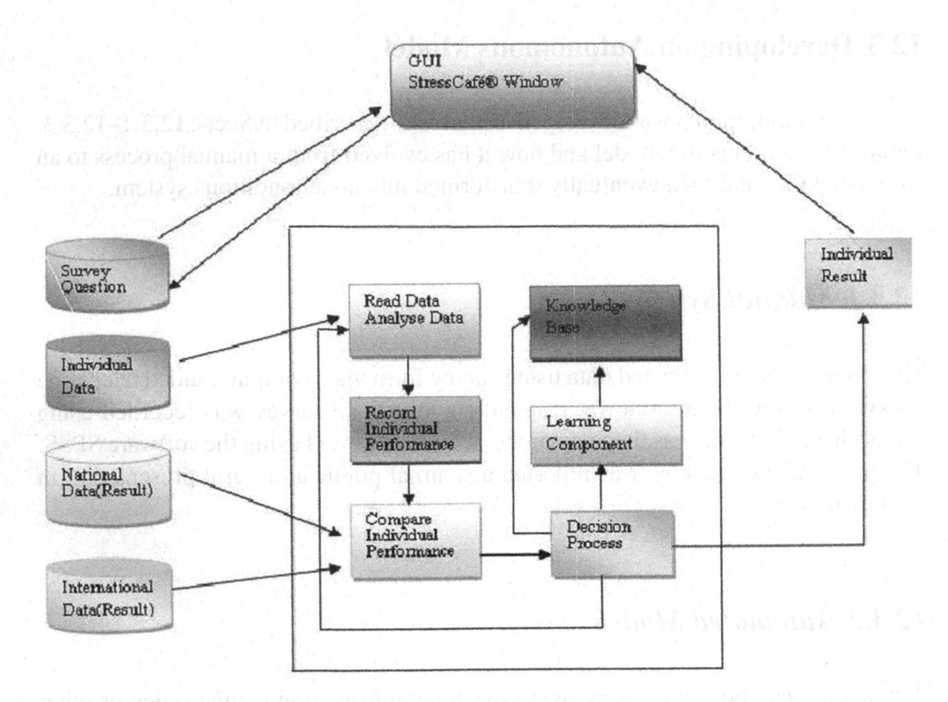

Fig. 12.3 Intelligent agent framework used in the StressCafé

The four databases are defined as:

Survey Database: The survey database descriptive is a list that stores the survey form and user information.The questionnaire contains a list of category based questions that address each element included in the survey questionnaire.

User or Individual Database: The user database records information about each user as they the user registers or undertake the survey.

National or Benchmark Database: The national database is a collection of all results from all surveys conducted nationally. For example: selection of an individual state and or industry in Australia.

International Database: The international database is similar to the national database. It contains survey results constructed in countries around the world. For example: countries outside Australia. The international database is not used for the purpose of benchmarking during this research, however this becomes a future option.

The remaining components include:

- The Decision Making capability which analyzes the data against the national benchmark results.

- The Knowledge Base capability that presently contains the domain knowledge, with rules from both psychology and AI .
- The Learning Component capability which adapts rules, updates knowledge, with every iteration of the system run in time.
- The Feedback capability that generates the result that are presented graphically via statistical analyses. The individual user receives a from of visual graph providing a visual recommendation, warning and/or alert to the user in real time.

The model was developed using Apache as server, Personal Home Page: Hypertext Preprocessor (PHP) as a scripting language, My Structured Query Language[2] (MySQL) database for storing, pre-processing and manipulating data. It is a fully automated model which is hosted within the e-portal Stresscafé. The automated interactive model can be used on-line, as a standalone system, in mobile equipment, kiosk to name a few of the possible application as it is web based and used distributed protocols. The model allows user to take the on-line survey and, receive feedback of their work related stress level in the form of a graph that dynamically benchmarks the users score against the national score [33].

12.3.3 Approaching Autonomy

To achieve autonomy, by autonomy it is meant that a system can operate independently with out any human interaction, the automated model was modified to include MAS architecture as shown in Fig. 12.6. Wooldridge and Jennings [7] state that an agent is a hardware and/or software based computer system which display autonomy, social adeptness, reactivity, and proactive. Using MAS over single agent reduces design complexities and is scalable. The transformed model with MAS architecture comprises of four agents which are individually programed and developed but these agents interact with each other in order to acheive mutual goals. When agents interact with other agent, it enables them to learn more about the environment in which they are situated and also use other agents intelligence and skills to solve problem efficiently. The design development and implementation of the autonomous model is described in Sect. 12.4.

12.4 Multi-Agent System Technology Incorporated to Automate the Intelligent Agent Framework

The IAF was transformed by incorporating multiple agent to operate in framework. A decision making system was incorporated MAS. The resulting system was called IMADA. The MAS framework provides IMADA with individual capabilities which

[2] My is the daughters' name of the MySQL co-founder, Michael Widenius, which he applied to this Relational DataBase Management System (RDBMS).

are the user agent and the benchmark agent. The agent manager and the agent facilitator coordinates the functioning of the different agents via message protocol. The transformation of the automated decision process within the IAF to IMADA is shown in Fig. 12.4.

The database agent within IMADA consists of user agent, benchmark agent as shown in Fig. 12.4, which have been used to maintain and update the tables (user table and Benchmark table) within the databases, as more and more user uses the system, the data from each user table is periodically added to the benchmark table formalizing the data. This is achieved using threshold values set by the administrator of the system. The data to be added to the benchmark table must be validated after removing outliers. The protocol from the user agent to the benchmark agent via the agent manager have been represented using the sequence diagram as shown in Fig. 12.5. MAS architecture is used to combine the user agent and Benchmark agent used to update and maintain each tables within the databases.

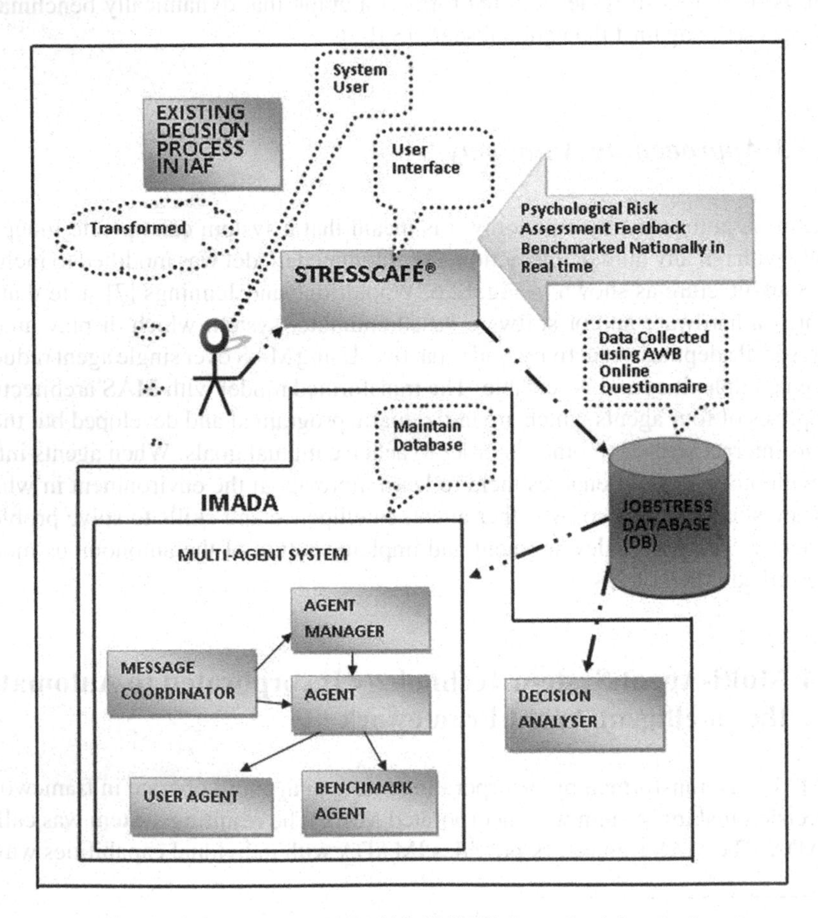

Fig. 12.4 The transformed process in IAF within IMADA for maintaining databases

The database updating process becomes tedious if the data had to be pre-processed and appended manually, especially when the interval is unknown and may also compromise efficiency and effectiveness of the system. Thus to optimize the performance of the required operations on the tables within the database, MAS architecture based design has been proposed, developed and implemented [34].

Now in order to further enhance the capabilities of the automated IAF to a autonomous model, the decision The decision process within IMADA is transformed as shown in Fig. 12.6 with more agent capabilities using MAS techniques to form a hybrid intelligent system.

To optimize the data analysis process, the IAF framework incorporates MAS linked to the analyzer IMADA. Within IMADA there are four component /elements. The four component/ element within the evolved decision analyzer are:

- Database capability,
- Knowledge capability,
- Neural Network capability, and
- Fuzzy Logic capability.

12.4.1 Database Capability

There are four discrete databases that are embodied as individual components. Each element is integrated to operate autonomously. The database management agent maintains and updates the user and the national benchmark database. Then a neural network that uses the back-propagation algorithm is being programmed using the programming language Java. Then using fuzzy logic, the output from the neural network is transformed in to linguistic grade. Problem solving, data analyzing, decision making are complex task that require techniques that are feasible, cost-effective and efficient. Traditional technique such as hard computing which include expert system and operational research; soft computing that includes: neural networks, fuzzy logic, and genetic algorithms. These techniques are mostly referred to as intelligent technique; these intelligent techniques have their own strength and weaknesses and cannot be applied to solve all problem. Thus integrating two or more technique to solve problem can provide better and precise solution. Thus the analyzing component has been developed using hybridized artificial intelligence technique: neural networks agent and a classifier agent. The results obtained using the new analyzer were at least as accurate and in some cases superior to previously analyzed results [32–34].

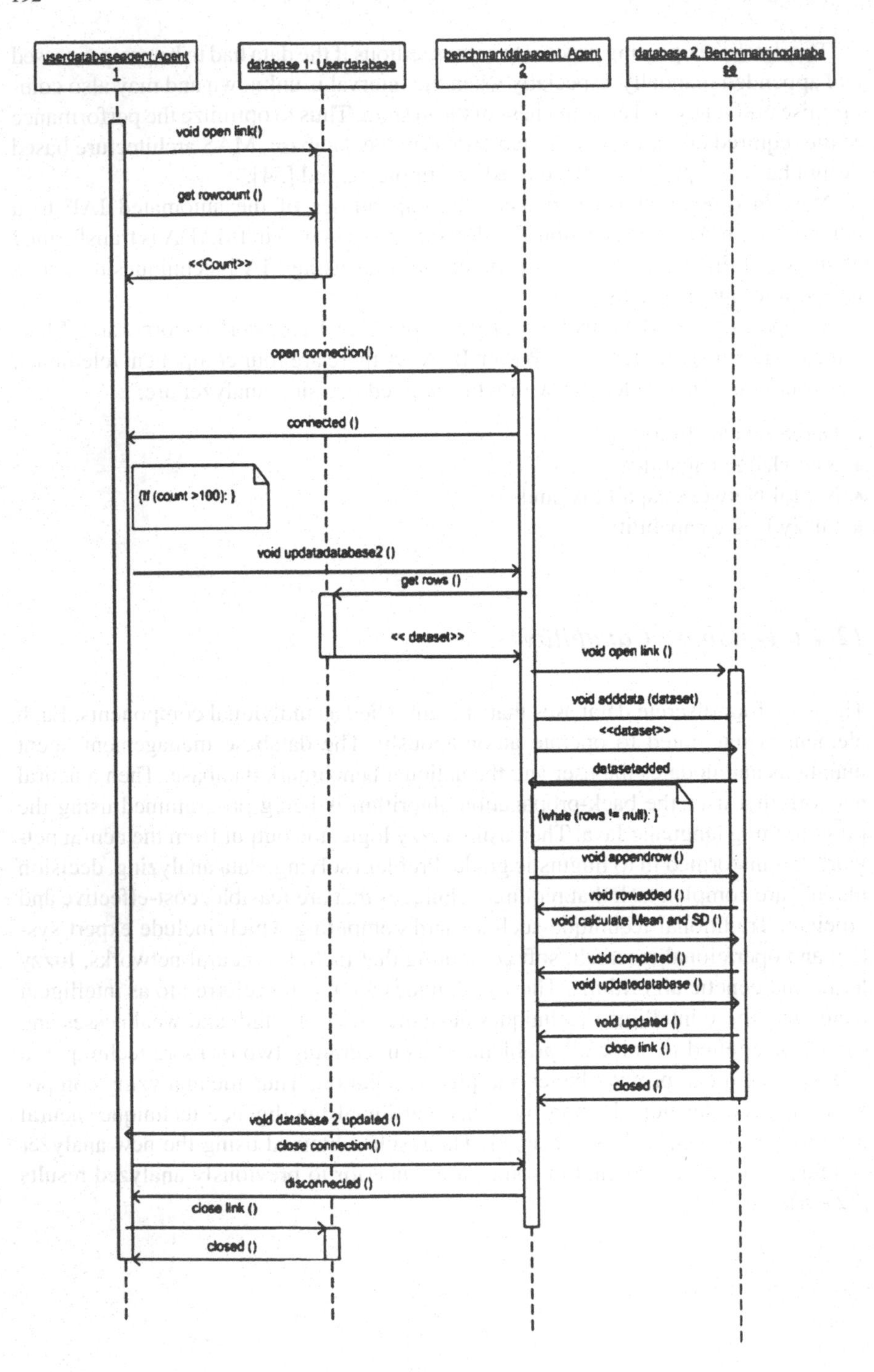

Fig. 12.5 Sequence diagram capturing agent protocol

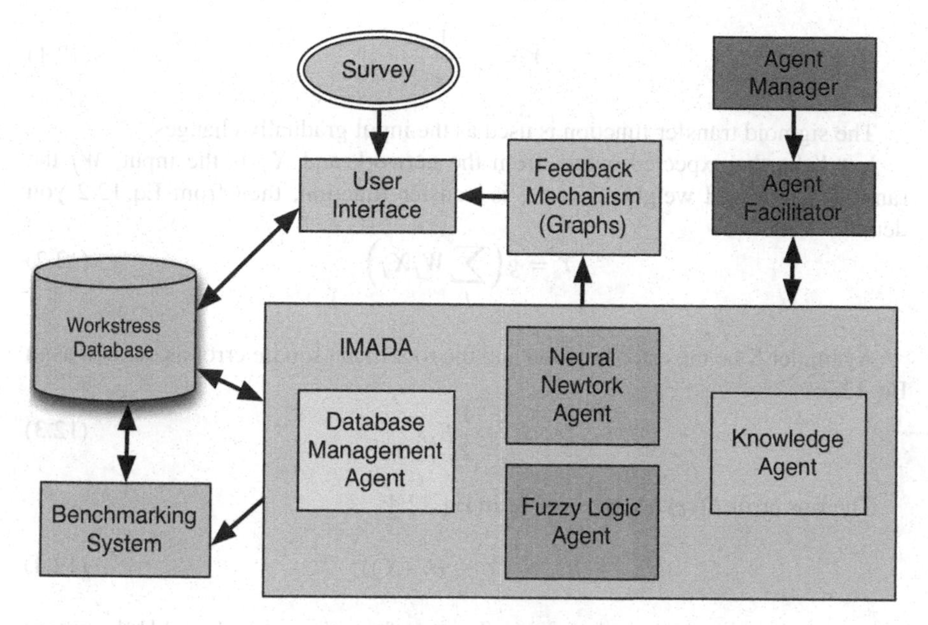

Fig. 12.6 Decision analyzer to include various agent capabilities

12.4.2 Knowledge Base Capability

The Knowledge base consists of facts and rules about the subject at hand, presently it contains knowledge from psychology, mostly relating to the work-stress survey questionnaire terminology. Domain knowledge along with rules from psychology and artificial intelligence are being added to the knowledge base.

12.4.3 Neural Network Capability

A Neural Network is a combination of a digital representation of a human inspired neuron, where the network is the means of creating the virtual brain. A technical neural network consists of simple processing units, the neurons, and directed, weighted connections between those neurons. The function of the neural network is to produce a pattern on its output which is correct for the class when an input is presented to the network.

The neural network agent is a neural network, which have been programmed using the back-propagation algorithm in Java. The network consists of five input layers, one hidden layer with six neurons to derive a single output. The weights are randomized in the beginning and the transfer function used is sigmoidal as given in Eq. 12.1, where Y is the output, x the input and e the exponential function.

$$Y = \frac{1}{1 - e^{-x}} \tag{12.1}$$

The sigmoid transfer function is used as the input gradually changes.

Let Y be the expected output from the network and X_j is the input, W_j the randomly assigned weight. Given g is transfer function, then from Eq. 12.2 you derive,

$$Y = g\Big(\sum_j W_j X_j\Big) \tag{12.2}$$

Again, let E be the error function and the root mean square error as defined as in Eq. 12.3:

$$E = \frac{1}{2} {E_{rr}}^2 \tag{12.3}$$

The rms error $E(rr)$ is expressed as in Eq. 12.4:

$$E_{rr} = (Y - H_w(X))^2 \tag{12.4}$$

where, $Y = true$ output or the expected output from the network and H the output from the network when the input X along with weight W presented to the network. H is given by Eq. 12.5:

$$H_w(X)) = g\Big(\sum_j W_j X_j\Big) \tag{12.5}$$

If the target output is same as the output from the network then it has reached the solution, but if the target output is not the same as expected then the error term is calculated using Eq. 12.4. After calculating the error, the error is minimized by taking the partial derivative of E with respect to W_j as in Eq. 12.6.

$$\frac{\partial E}{\partial W_j} = -E_{rr} \times g'(in) \times X_j \tag{12.6}$$

which gives the rate of change of error with weights. The updating rule is then given by Eq. 12.7

$$W_j \leftarrow W_j + \alpha E_{rr} \times g'(in) X_j \tag{12.7}$$

α, denote the learning rate. The updating rule for a multilayer network is calculated using Eq. 12.8:

$$W_{k,j} \leftarrow W_{k,j} + \alpha \times a_k \times \Delta_j \tag{12.8}$$

given:

$$\Delta_j = E_{rr} \times g'(in_i).and \Delta_j = g'(in_j) \sum_j W_{j,i} \Delta_i \tag{12.9}$$

Thus, Eq. 12.8 serves as the update rule when the error is propagated to the hidden layer via the output layer. The calculated error information is fed back to the system which makes all adjustments to their weight for each training set presented to the network. This process is repeated until the desired output is acceptable.

12.4.4 Fuzzy Logic Capability

The fuzzy logic agent transforms the crisp values into grades of membership for linguistic terms. By using the fuzzy logic agent the crisp output from the neural network is mapped via membership function to a human readable and presentable lingustic form.

A fuzzy set A is defined by a set or ordered pairs, a binary relation, where, A(x) is a function called membership function; A(x) specifies the grade or degree to which any element x in A belongs to the fuzzy set A as shown in Eq. 12.10:

$$A = (x_1 \mu_a(x)) | x \varepsilon A, \mu_a \varepsilon [0, 1] \tag{12.10}$$

where, each element x in A is a real number A(x) in the interval [0, 1] which is assigned to x. Larger values of A(x) indicate higher degrees of membership [29].

A work-flow diagram for the fuzzy logic agent is given in Fig. 12.7.

The input range is [0–6] The input status words are:

1. Above,
2. Slightly above,
3. Just right,
4. Slight below, and
5. Below.

The output action in words is defined as:

1. Very high,
2. High,
3. Medium,
4. Low, and
5. Very low.

A sample of the rule-base is listed as follows:

1. *If (the user score is above 5) and (the benchmark score is less than 5). Then user level of stress is very high.*
2. *If (the user score is below 4) and (the benchmark score is less than 5.) Then user level of stress is high.*
3. *If (the user score is just right) and (the benchmark score is just right). Then user level of stress is medium.*

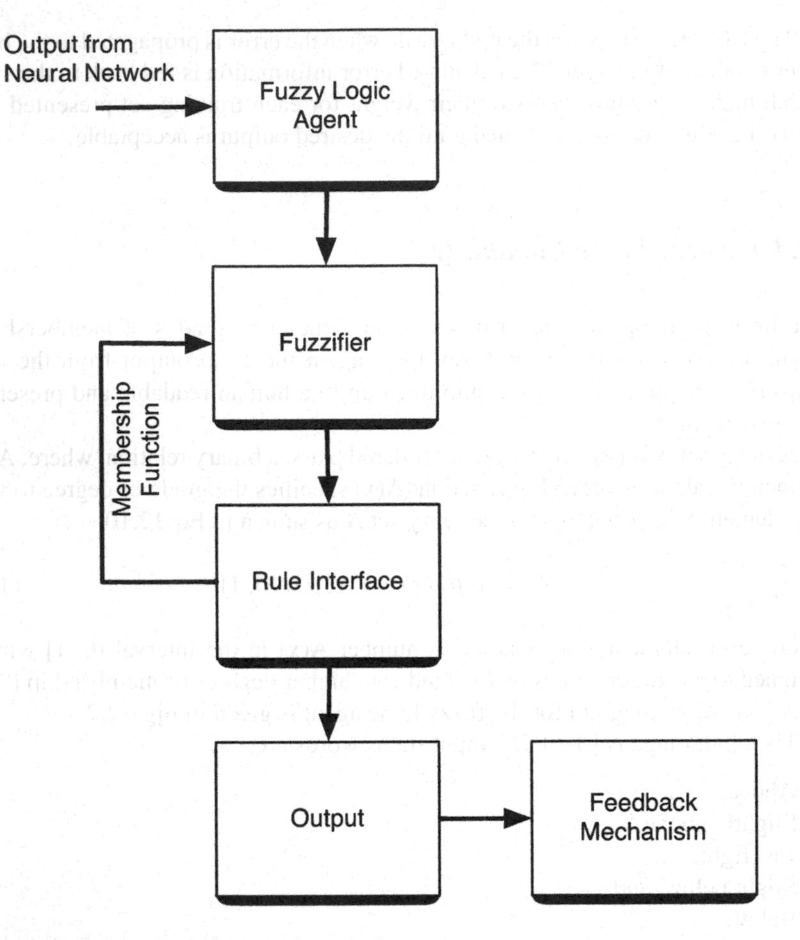

Fig. 12.7 Work-flow diagram for the fuzzy logic agent

4. *If (the user score is above 3) and (the benchmark score is less than 3). Then user level of stress is low.*
5. *If (the user score is below 2) and (the benchmark score is above 2). Then user level of stress is very low.*

12.5 Case Study

The different intelligent technique are individually programmed and then integrated and implemented. The model is implemented and tested on-line as well as off-line using the e-portal StressCafé. The StressCafé is an e-portal that hosts on-line surveys and provides e-feedback to aid the translation of research into policy and practice, also it is intended that the website will provide e-therapy and e-counseling along with the nationally significant workplace surveys. This is a one stop web-shop which can be accessed by industry, individuals, government bodies, communities to collect, compare and share information in relation to work related psychological risk [32–34] Within the StressCafé the hybrid intelligent model is implemented and tested using data collected via survey from six states and territories within Australia.

The data collection commenced in the year 2009, completing four waves of collection by the year 2011. The present study uses data that has been collected from four Australian States which are North South Wales (NSW), Western Australia (WA), South Australia (SA) and Tasmania (TAS). Data was also collected from two Australian Territories namelyAustralian Capital Teritorry (ACT) and Northern Territory (NT) [1, 35].

12.5.1 Case Study with Special Emphasis on Depression

The collected data was preprocessed and stored by the database agents as these data were collected from various states and territories across Australia; they will be used to benchmark individual user score nationally. The Benchmarking database is also being updated with user data based on a threshold value which was 50 in this case. As user takes the on-line survey, their data is preprocessed and analyzed by the neural network agent. As mentioned in Sect. 12.4.3 a backpropagation neural network is used. The neural network takes five parameter input: *Industry, User Mean, Total Mean, Mean+2SD, Mean-1SD* which are chosen as the user level of work stress will be benchmarked based on the industry they work. The user mean for a particular section of the questionnaire (depression in this case) is calculated. The Total mean is the mean calculated for the same section of the questionnaire data collected nationally. The standard deviation (SD) for the user mean for depression is calculated [32]. The network consists of:

1. Inputs: 5
2. Desired Output: 1 O (the output from the network), one output
3. Weight = W, the weights are assigned randomly.
4. Learning rate = η
5. Hidden neuron = 6
6. Hidden layer = 1

The sigmoid transfer function used is as given in Eq. 12.1, which is considered as it is the best optimizer for this case. The output from the neural network as can be seen from Table 12.1 is in a numeric form or crisp form, which is processed by Fuzzy logic agent.

The fuzzy logic agent transforms the crisp values into grades of membership function for linguistic terms, of fuzzy sets as shown in Table 12.2. The fuzzy output is stored and presented to the uses via the feedback mechanism.

Table 12.1 Input and desired output from the neural network

Inputs					Output
Q8B	Mean total	Mean+2SD	Mean-1SD	Depression per user	Target output
4	2.7368	9.6134	−0.7014	13.6060	5
10	3.8433	11.8533	−0.1616	4	4
13	3.6797	0.0412	−0.0509	6	5
10	3.8433	11.8533	−0.1616	1	2

Table 12.2 Desired output from the neural network and fuzzy logic

Inputs			Outputs		
Q8B	Mean total	Depression per user	Target output	Network output	Fuzzy output
4	2.7368	13.6060	5	4.4505	High
10	3.8433	4	4	4	High
13	3.6797	6	5	6	High
10	3.8433	1	2	1	Low

1. very high,
2. high,
3. medium,
4. low, and
5. very low.

From the summary provided in Fig. 12.8 the bar graph represents the user score in black where as the grey score represents nationally benchmarked data on a 0–100 scale for Depression in this instance. The summary also depicts user response along with the average population response (Benchmarked) that a particular user is at that particular point in time.

12.6 Conclusion

This chapter presents the evolution of a system development from manual to automatic through to the autonomous stage. The system design, development and implementation have been presented in this paper. This hybridized autonomous tool that was developed using MAS techniques have been applied to solve a problem in the field of psychology. The problem was to "analyse work-stress data as the user participates in an on-line survey and submits his or her responses, providing feedback to the user in real-time". The model uses the e-portal the StressCafé, but can be hosted on-line, as a standalone system, in Kiosk to name a few. The e-portal StressCafé, hosts on-line surveys and provides e-feedback to aid the translation of research into policy and practice, also it is intended that the website will provide e-therapy

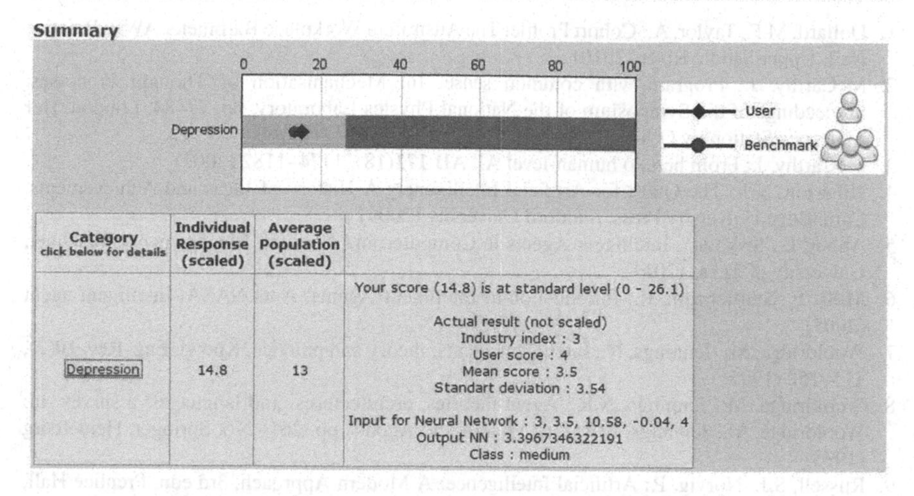

Fig. 12.8 Benchmarked user feedback summary

and e-counselling along with the nationally significant workplace surveys. Work stress related data collected from four states and two territories across Australia has been used to test the model developed. A feedback is provides in the form of graph to participant who completes a work-based psychological risk assessment survey by comparing individual results to nationally benchmarked score. The autonomous hybridized model processes data automatically optimizing, productivity and predictability.

12.7 Future Work

Future work could include an emotion recognizer agent to record user emotions, thus adding another dimension, enabling more in-depth analysis of work stress related data. Also the knowledge base could be developed further. A learning component would be another functionality of the model which can adapt to the dynamic environment and learn in real time.

Acknowledgments This research work is supported by the Australian Research Council Linkage Grant LP 100100449. The authors would like to thank my Prof. Maureen Dollard, Director, Centre for Applied Psychological Research for her support and encouragement and also would like to thank research assistant Wayan Firdaus Mahamoody for his assistance and co-operation in developing the StressCaf Website.

References

1. Dollard, M.F., Taylor, A.: Cohort Profile: The Australian Workplace Barometer AWB. Prentice Hall, Upper Saddle River (2010)
2. McCarthy, J.: Program with common sense. In: Mechanisation of Thought Processes, Proceedings of the Symposium of the National Physics Laboratory, pp. 77–84. London, Her Majesty's Stationery Office (1959)
3. McCarthy, J.: From here to human-level AI. AIJ **171**(18), 1174–1182 (2007)
4. Nilission, N.J.: The Quest for Artificial Intelligence: A History of Ideas and Achievements. Cambridge University Press, Stanford University (2009)
5. Abbot, L., Siskovic., Intelligent Agents in Computer and Network Managements. Teachnet, University of Texas (2011)
6. Mills, F., Stufflebeam, R.: Introduction to Intelligent Agents. Ants NASA, Intelligent agent (2005)
7. Wooldridge, M., Jennings, N.: Intelligent agents: theory and practice. Knowl. Eng. Rev. **10**(2), 115–152 (1995)
8. Wooldridge, M., Jennings, N.R.: Agent theories, architectures, and languages: a survey. In: Wooldridge, M., Jennings, N.R. (eds.) Intelligent Agents, pp. 261–276, Springer, Heidelberg (1995)
9. Russell, S.J., Norvig, P.: Artificial Intelligence: A Modern Approach. 3rd edn. Prentice Hall, Inc., Englewood Cliffs (2010)
10. Tweedale, J., Jain, L.C.: Embedded automation in human-agent environment, Adaptation, Learning and Optimization, vol. 10. Springer, Berlin (2011)
11. McArthur, S.D.J., Davidson, E.M.: Concepts and Approaches in Multi-Agent Systems for Power Applications. Intelligent Systems Application to Power Systems (2005)
12. Nakano, K., Yamaguchi, D., Katayama, F., Takahashi, M.: The Medical Diagnosis Support System with Intelligent Multi Agent Techniques by Performance Differential Difference. In: Fifth International Workshop on Computational intelligence and Applications, IEEE SMC Hiroshima Chapter, pp. 10–12, Hiroshima University, Japan, Nov 2009
13. Tweedale, J., Ichalkaranje, N., Sioutis, C., Jarvis, B., Consoli, A., Phillips-Wren, G.: Innovations in multi-agent systems. J. Netw. Comput. Appl. **30**(3), 1089–1115 (2007)
14. Tweedale, J.W.: Autonomous agent teaming providing enhanced communication components. Phd Thesis, University of South Australia, Australia (2009)
15. Mangina, E.: Intelligent agent-based monitoring platform for applications in engineering. Int. J. Comput. Sci. Appl. **2**(1), 38–48 (2005)
16. Jung, I., Thapa, D., Wang, G.N.: Intelligent Agent based Graphic User Interface (GUI) for e-physician. World Academy of Science, Engineering and Technology, vol. 36 (2007)
17. Agah, A., Tanie, K.: Intelligent graphical user interface design utilizing multiple fuzzy agents. Interact. Comput. **12**, 529–542 (2000)
18. Zdrenghea, V., Man, D.O., Tosa-Abrudan, M.: Fuzzy Clustering in an Intelligent Agent for Diagnosis Establishment. Studia Universita, Babes-bolyai, Informatica, vol. 55, issue 2 (2010)
19. Moreno, A.: Medical Applications of Multi-Agent Systems. Computer Medical Applications, vol. 3, 2002
20. Warren, M., Walter P.: A logical calculus of ideas immanent in nervous activity. Bulletin of Mathematical Biophysics, vol. 5, pp. 115–133 (1943)
21. Rosenblatt, F.: The perceptron: a probabilistic model for information storage and organisation in the brain. Psychology Review, vol. 65, pp. 42–99 (1958)
22. Rumelhart, D., Hilton, G., Williams, R.: Learning representations by backpropogation errors. Nature **323**, 533–536 (1986)
23. Patterson, D.W.: Artificial Neural Networks Theory and Applications. Prentice Hall International pp. 247–264 (1996)
24. Rojas, R.: Neural networks-A systematic introduction. University of Pennsylvania Law Review, **154**(3), chaps 2–6 (1996)

25. Werbos, P.J.: Beyond regression: new tools for prediction and analysis in the behavior science. PhD. Thesis, Harvard University, Cambridge, pp. 99–154 (1974)
26. Parker, D.: Learning Logic, Tecnical report Tr-47, Center for Computational Research in Economics and Management Science, MIT, Cambridge, pp. 99–154 (1985)
27. Rumelhart, D.E., McClelland, J.L., and the PDP Research Group. Parallel Distributed Processing (PDP): Exploration in the Micro-structure of Cognition, vol. 1, The MIT Press, Cambridge, pp. (76, 77, 154, 155, 219, 329) (1986)
28. Tweedale, J.: Fuzzy control loop in an autonomous landing system for unmanned air vehicles. In: IEEE International Conference on Fuzzy Systems (FUZZ-IEEE), pp. 1–8. IEEE Press, Barcelona (2012)
29. Zadeh, L.A.: Fuzzy sets. Inform and Control **8**, 338–353 (1965)
30. Jang, J.S.R., Sun, C.T., Mizutani, E.: Neuro-Fuzzy and Soft Computing. A computational approach to learning and machine intelligence, Mathlab Curriculum Series (1997)
31. Zang, Z., Zang, C.: Agent-Based Hybrid Intelligent Systems. LANI 2938, pp. 3–11, Springer, Berlin (2004)
32. Ghosh, A., Nafalski, A., Tweedale, J., Dollard, M.: Hybridized technique to analyse work-stress related data via the StressCaf, ICCESSE 2012: International Conference on Computer, Electrical, and Systems Sciences, and Engineering. Bali, Indonesia, Oct 24–25, (2012)
33. Ghosh, A., Nafalski, A., Tweedale, J., Dollard, M.: Using hybridized techniques to develop an online workplace risk assessment tool. J. Inform. Control Meas. Econ. Environ. Prot. **4b**, 42–45 (2012)
34. Ghosh, A., Tweedale, J., Nafalski, A., Dollard, M.: Multi-agent based system for analysing stress using the Stresscafe. Advances in Knowledge-Based and Intelligent Information and Engineering Systems, Amsterdam, vol. 243, pp. 1656–1665 (Sep 2012)
35. Taylor, A., Gill, T.: Eleonora. Population Research and Outcome Studies, University of Adelaide, Adelaide, Australia, Dall Grande (2009)

Index

J. W. Tweedale and L. C. Jain (eds.), *Recent Advances in Knowledge-based Paradigms and Applications*, Advances in Intelligent Systems and Computing 234,
DOI: 10.1007/978-3-319-01649-8,